HEAT

BY

JACK FRERKER

*To Marsha & Shawn —
With my love —
Fr. Jack*

PAX PUBLICATIONS of OLYMPIA, WASHINGTON

Published by PAX Publications
7315 Henderson CT SE, Olympia WA

Copyright © 2003 by Jack Frerker. All rights reserved. No part of this book may be reproduced, stored in a retrieval system, or transmitted, in any form or by any means, electronic, mechanical, photocopying, recording, or otherwise, without the express written permission of the author.

Printed in the United States of America
by Bang Printing, Brainerd MN

This is a work of fiction, and while based on experience, all names, characters and incidents are from the author's imagination or used fictitiously. Reference to real persons is not intended and should not be inferred.

LIBRARY OF CONGRESS CATALOGING-IN-PUBLICATION DATA
Library of Congress Control Number: 2003092940
Frerker, Jack, 1937 –
    HEAT
    ISBN 0-9740080-0-1

ACKNOWLEDGMENTS

To George Goodin and Ferd & Joan Potthast for their editorial suggestions, and to Garn Turner for his lovely cover, my deep thanks. Thanks also to Mark Bersano and Richard Swanson who turned the text into printable copy, and to LaVerne O'Brien and Frank & Julia Ladner, for assistance with publishing this book.

Most of all, my everlasting gratitude to my teachers: Paul Pichotta and Robert Eimer. They introduced me to the vast and beautiful world of the story, and I've been entranced ever since!

Long-standing friends all, like many others who are here nameless, they have encouraged my creativity in way. Creating a fictional world is satisfying, but it's also a curious endeavor much in need of sustenance. My friends' encouragement, more than anything else, has provided that. I cannot thank them enough!

And finally to you readers, my gratitude for allowing a small glimpse of life from the region of my origins into your own lives. My story reflects *something* of that area, but because it is fiction, I hope you understand that Southern Illinoisans are really only partly like they appear here. Their actual lives are more rich and complex than my characters'. I hope this glimpse encourages you to meet some of them in the flesh.

Jack Freiker

CHAPTER I

Southern Illinoisans owe it to everyone, themselves especially, to know what's really going on beneath the surface, behind closed doors, even in peoples' thoughts. And if at first you're not on the money about the thoughts, keep at it and soon enough you'll get that too! Part challenge, part thrill, mostly it's how to keep things in balance. That's important for understanding someone like John Henry Wintermann. He knows his town and parish, and jokes about that resulting from hearing confessions. But he's not fooling anyone or even trying to. Nonetheless, don't misunderstand: natives aren't offended at that kind of knowledge *or* the quest for it, because they all practice the art. And pretty well, at that. Some are experts, like Father John, which is fortunate for Algoma. But that's getting ahead of things.

Small Southern Illinois towns are so alike only natives can distinguish Algoma, John Wintermann's town and the place where it happened, from neighboring hamlets. A county seat -- that narrows the field -- with three churches, all Mainline. Which narrows it even more. Hard to say why, but Algoma has no Bible churches. Beyond these, there are simply no other distinctive characteristics obvious to casual observers. That is, to outsiders.

They're a rarity in Algoma, outsiders are. With Interstates twenty miles or so away in three directions and nothing much for tourists, Algoma's outside most traffic patterns unless you're heading there directly, something only grown children do several times a year: Christmas, and Memorial Day or Labor Day, and perhaps again on Thanksgiving or Independence Day -- or *Decoration Day* and *the Fourth,* as those summer holidays are known locally.

Things haven't changed in decades, since maybe as far back as when they laid down the hard road! So those grown-up kids say. Not true, of course. It's more a measure of how completely they've emigrated. They don't notice a new name over the teen hangout, or what was the hangout when they hung out there. Now mainly a drugstore again, its soda fountain is still operational. Or church signs. Not cute Protestant ones -- *Seven days without prayer make one weak* -- but those that trumpet pastors' names. They shift every decade or so. At least the Protestants' do. Or the water tower's new paint job. *Lots* of changes! The population is noticeably older, and with fewer children, lawns are neater. Things like that! Things Father John tends to notice, because he's a native.

Of the region, mind you. He grew up in Maple Grove, not that it matters. Algoma and the Grove are a lot alike, with streets named for trees and Presidents and, as in most neighboring towns, one named *Front.* Except for a courthouse! Maple Grove hasn't got one. But they're enough alike so's you can understand the basics once you move in. Just learn names. And relationships -- to avoid offending someone's kin, which could easily be half the town!

Father John's been around long enough to be well beyond all that and into local history. Fact is, he's beyond that too. Twenty-eight years already, he'll surely be around a bunch longer -- probably die there, unless he retires first, which isn't likely. At sixty-three it's a pretty good bet he'll stay, unless the bishop gets mad at him or orders a general house cleaning, something the seventy-five or so priests would almost certainly resist anyway. Just common sense he'll be there the duration. Which provides even more incentive to *really* get to know everyone: something die-hard, down-home, never-any-place-but-Southern-Illinois natives consider their *duty*. That's certainly the way Father John sees it. And, of course, as a native he's good at it -- so good, he might just as well be from Algoma in the first place. By now, only the oldest residents realize he isn't.

So, first off: Algoma's small, very much *Southern* Illinois, and improbably half Catholic. And Father John's not only a Southern Illinois native, for all practical purposes he's an Algoma native. Not much there escapes his notice. But now, to city people this makes him as nosy as a mother-in-law. City folks don't really understand the duty part. Maybe they just don't have time for that. Or take it.

Incidentally, a *city* is any place big enough to hold six or seven little towns. Peoria, certainly, and Springfield; Bloomington, Champaign and Rockford; and, of course, Chicago. But lots of other places too not as big, like Decatur or even Pekin. But then, none of those are in *Southern* Illinois. Springfield likes to think it is; or at least Chicagoans think it is. But *natives* understand that *Southern* Illinois begins *maybe* at the I-70 parallel. Certainly at Mount Vernon! Everything else is *north*. And it's northerners and city folks who just don't grasp the duty part.

Well then, you can imagine his consternation when Annie Verden turned up unexpectedly dead with no history of illness or signs of foul play. In Southern Illinois *unexpected* means tornadoes, earthquakes and little bitty Cobden nearly taking the state basketball title. It doesn't include sudden death sans obvious probable cause, such as drunk-driving or truck stop food poisoning.

Father John's consternation is especially noteworthy since he's always trying to unravel puzzlements, sniffing out clues where even the FBI wouldn't think to look. That is to say, his is an active imagination, something decidedly helpful with the duty part of staying on top of small town happenings. Not that he initially suspected anything. He simply went the six steamy miles to Burger on an early September morning for an anointing. Turned out to be Annie, and despite the weather, no longer warm at that. So the sacrament was more comfort than last-minute express ticket to heaven. But since Annie had few relatives anyway and none of the acknowledged ones nearby, that day's was a lonely, thankless task. Routine, even.

Her domestic had found her body. Obviously deceased, standard county protocol dictated she be taken to the only county medical facility, Saint Luke's, for a death certificate and possible autopsy. Since Annie was on in years, dying in sleep of old age seemed reasonable. And absent anything suspicious, no autopsy was ordered. Though, so far as anyone later noted, and her doctor verified, she'd been healthy as a mule. And easily as stubborn, everyone knew.

Something Father John noticed at the hospital got him thinking later that day. He made little of it at the time and wouldn't even have remembered later but for a series of apparently unrelated things. Only an imagination like his would have pursued it, anyhow. You dasn't say there's nothing to do in small town USA! It's precisely what city folks style *having nothing to do* that allowed Father Wintermann to develop his imagination in the first place. And it gave him incentive to keep puzzling over Annie's death.

CHAPTER II

It all started a couple of months before Labor Day at Saint Helena's church, long before Annie was a corpse, while things were slowed to a crawl even by Algoma standards -- during the sticky, dog days of a Southern Illinois summer. You know the kind of weather: triple digit temperatures and nearly triple digit humidity. Farmers crave it and locals brag on it, though few actually relish it, till they can mention it, casually, to an outsider. It wouldn't be summer without lots of that. When you have weeks of it on end, it's a Southern Illinois summer. That year it was a *real* Southern Illinois summer, since it had been going on over a month already and was to continue a good while longer still.

Since Father John fares no better in that weather than anyone else but has to wear more stuff, at least at Mass, he was into his *summer liturgical mode.* That's when from start to finish, with a hymn or two and sermon, the whole Mass takes about twenty-six minutes, even with a second collection and announcements. It could be different with air-conditioning. But his people aren't keen on spending that kind of money. And they *prefer* short Masses, he knows.

Anyway, it was that kind of summer. Animals move slowly and stay shaded whenever possible. Only teens, the mad dogs and Englishmen of an American summer, can be found working on their tans. And you've got to dig down a foot or more through dry, hot topsoil to find fishing worms. The weather only broke as they were laying the duchess in the ground, so it was just a touch over *three* months solid that the region simmered that year -- a factor in the case, actually.

It was at one of those Blitzmesse that Father John preached his *Summer Sermon*. Parishioners look forward to it as a seasonal marker as well as a model of brevity and precision. *Today it is as hot as hell. Think about that! In the name of the Father, and of the Son, and of the Holy Ghost.*

That Sunday was the first time two of Annie's out-of-town relatives heard it. They commented on it after Mass. Favorably. By now a usually annual affair, the sermon *is* appreciated by parishioners but rarely acknowledged any more. The strangers acknowledged it, however. Which impressed the priest; massaged his ego. For a brief moment he saw himself as the fiery, charismatic preacher he had at times aspired to be. *Nice folks. Very. Discerning too.*

The two were visiting their aunt on the cusp of June/July, an odd time. Nothing special going on; no holidays for a week or so. You know: no parades or street fairs! Not even a Methodist Ice Cream Social. Too hot. Can't keep the homemade stuff from souping up. Wheat was near harvest. Strawberries were long gone, and most lettuce. Sweet corn and tomatoes weren't in yet. Zucchini, onions and radishes were all you could get from the garden, plus some cucumbers. He remembered thinking that Annie, who still cooked, and pretty well at that, wouldn't have anything special except maybe canned stuff. Zucchini just isn't haute cuisine no matter how it's served. It's just not how to extend a real Southern Illinois welcome. Annie saw it that way too, he knew.

But then, these fellows didn't strike him as the sort you'd worry about pleasing at table. Kind of thin and scrawny, both of them -- not the kind to eat their way through summer, from the asparagus and strawberries all the way through melons and the late corn, plus everything in between, including the beef and pork. Not like natives, that's for sure. It's one way to identify outsiders, in fact: they often look slightly peaked, and their pants hang a little limp like, sliding down over their hip bones. They're not *corn-fed,* which in Southern Illinois applies to more than livestock. Southern Illinois mothers consider it abnormal not to fuss till you clean off your plate and the table to boot.

So, these two kind of stood out. The only question was whose visitors they were. Since Annie had trouble getting around and wasn't often at church, especially in the heat, they weren't immediately identifiable, sitting by themselves in one of the few unclaimed front pews. Father John thought it either very odd or very nice -- he'd determine which afterwards -- that two male strangers came to Mass. He cleared it up later when he shook hands at the front door and held on long enough to get what he wanted: Annie's kin, nephews -- great-nephews, actually -- brothers, thirtyish and unmarried, city folks and college professors besides! He was good at that sort of thing. Had it down to a question or two and several innocent comments designed to elicit answers without sounding like cross-examination. That, plus a bulldog grip which, while unimpressive to natives, never failed to get anyone else's attention.

In their case he decided it was nice, even though few college professors come to Algoma, especially any with Italian faces and

German names. After introductions, they smiled and complimented him on his *piquant amalgam of theological and topical material.* He decided they had at least learned couth in college. When it became clear they were serious, he was tempted to actually like them. But native that he was, he simply oozed charm and withheld anything near a full dose of affection. Still, they *were* discerning and appealing lads. Upon learning they'd be staying several more days, he announced he'd come to the Verden mansion for breakfast the next morning after Mass.

That's what everyone called it. *the mansion.* Once an impressive, three-story, many roomed and elegantly columned home atop the town's central hill, the Verden place now stood in need of at least one coat of paint and some generally younger hands to manage its restoration to former grandeur. Annie had let it slip, everyone agreed. But it was still imposing on a street of six or seven other imposing edifices. It was generally believed that Annie could easily manage the expense of its sorely needed refurbishing, if only she could be persuaded of its necessity. Town wisdom decreed that it wasn't her continued immersion in the dreamy past when that hill's homes were at the shining center of Algoma society, but more probably failing eyesight that allowed her to see elegance still glittering there, enough so, at any rate, to let her coast along -- in her very late seventies, would that be? -- without tending to such bothersome chores. She found it far easier to grace the front porch swing, hand fan and fly swatter at the ready, and preside over yet another summer as the minor nobility she'd been assured she was from girlhood.

There was more to be learned about her guests, and Father John had numerous times before shamelessly used lesser excuses to get homemade biscuits and gravy. Accordingly, on Monday he accompanied the young men up the hill to Annie's, for they had piously attended morning Mass.

The heat was already shimmering off road and sidewalk at a tad past eight thirty-five when they arrived, the smell of hot biscuits greeting them at the front door ahead of Annie's distant voice and well ahead of Annie herself. Barging in on someone like Annie was unthinkable, even though two of the bargers *were* houseguests, and relatives besides. Soon enough she appeared to unlatch the screen, a minor obstacle to entry but one that served in a small town to protect space if not person -- the only acceptable reason for hooking a flimsy screen door in an entire town of unlocked buildings, and the perquisite of only certain of the elderly at that!

The interior was an old woman's fantasy world. Immaculately clean, it was virtually stuffed with antiques, most of which looked brittle and directly challenging to the likes of Father John's frame. A tall man, and large as well, he carried his weight gracefully enough, but years earlier he had learned to distrust any except the amplest and sturdiest of chairs. Thus, he'd long since come to admire most of Annie's furniture from a distance. Besides antiques, there were also handcrafted things, many of Annie's own making, frilly and lacy, and leaving the priest unenthused yet assured they belonged there.

The smell of breakfast mingled with her thin, cheery greeting. "Come in, Father Pastor. Isn't this dreadful weather! I believe you're already well-acquainted with my nephews." She was a thin, now slight woman, who was wearing her *charming* face that morning.

"Yes, it *is* awful," he said as he stepped into the relative cool of the high-ceilinged front room and wiped his forehead with a large white handkerchief. "But I wouldn't say *well* acquainted, Miss Verden," he said, expecting the inevitable permission to call her *Annie,* which came almost before his sentence finished. "We only met yesterday after church, and just now again as we walked here. Tell me, how did you come by such erudite relations, Annie?" He stressed her name pointedly.

"Oh, don't sound so surprised that there's learning in my family, Father. These are my late sister's grandchildren. She didn't live here. Neither did her children, you probably remember. They grew up out East, don't you know, and both went to college. These boys too. Just never graduated, so to speak." She smiled at her own whimsy, as did the young men. Dutifully, it seemed.

Breakfast was leisurely and delicious. Besides freshly squeezed orange juice and biscuits with gravy, there were sausage links and extra biscuits for her homemade black raspberry jelly. With conversation over coffee it took nearly an hour, all on the summer porch, as the often screened-in back porches of such homes are called. The addition of a ceiling fan freed Annie's hands to present the meal graciously without the distraction of swatters and hand fans. That, and the fact that the porch was directly off the kitchen, made it ideal for breakfast or early evening meals, even in the hottest times.

There was never any question of installing a fan on the elegant ell-shaped front porch. Near enough to Annie, it might conflict with the swing's trajectory. Further away, it wouldn't cool a thing. Anyway, absolutely anyone can see how entirely out of place such a thing would be on the front porch of a fine old home. The seclusion of the summer porch, however, allowed Annie to accede gracefully and without social scandal to her sister's suggestion of nearly a decade earlier to install a *porch* fan. Still, she rarely used the area except to entertain, and then only over food of a morning or evening. Holding court was definitely a front porch affair. Not that it couldn't be done on the summer porch, but there it had to be squeezed in between trips to the kitchen and much clucking about eating more home cooking. As such, it could never be a full-fledged and leisurely court ceremony. That was reserved for long summer evenings on the front porch, from the swing that was her throne. Where the neighborhood could properly observe, besides.

Mind you, nothing like the whole town would troop past. But those who did paid homage. Even if they didn't intend to or weren't aware as they did so. Especially if they weren't aware! Annie was good at exacting homage. It was no longer a challenge. By her upper years it had become second nature.

Were someone so gauche as to intimate that's what she was up to, she'd demur and deny and soon demand apologies for daring to think it. Nonetheless, should one of a dwindling number of ladies of her generation even hint at the subject, it was a challenge instantly accepted with ease and nonchalance. And soon enough, Annie would elicit some sort of obeisance. In such moments a keen observer could detect the armor beneath her lace and the steel in her personality, both of which belied the genteel, fragile southern belle Annie took pains to appear. The delicacy of speech and wispy, aging body fooled outsiders, even some natives. But under the facade was someone not to be tampered with. In fact, look closely and one could discover a willful, opinionated and cunning woman, deliberate, extremely calculating, and even, as Father John had come to believe, on occasion downright mean.

The coffee ritual finale was the moment the priest began his work in earnest. That's not to say he was unobservant as they dutifully put away the huge breakfast. But over coffee he hauled out all the tools of his trade, stirred the heavy equipment and really began to dig. If Annie knew how to preside in courtly fashion, John Wintermann knew how to extract information.

"So, where'd you go to college?" he asked the older brother.

"Undergrad? Fordham. Brooklyn Catholics, it seemed the thing to do. Not that either of us regret it." His brother was nodding in agreement. "Fordham gave us a good education and was a fine experience. Jesuits, you know."

Whatever that means, Father John wondered, schooled as he had been by Benedictines.

Anticipating the next question, Joseph added: "We both got a B. S. in Chemistry. Then I went into Biochem at Wash U., while Tony ... "

Father Wintermann cut in: "In Saint Louis?"

"Yes."

"That's near enough I would have supposed you'd visited your aunt then. But I don't remember meeting you before."

"Actually, I was here, but only two or three times. I doubt you'd connect me with the college kid I was then. I've lost a bit of weight and a lot of hair. Anyway, that was well over ten years ago."

Father Wintermann was disappointed at not remembering *anyone* of college age visiting ten to fifteen years earlier. Annie didn't contradict it, so it must have slipped past him, perhaps the visits coinciding with his vacations. Still, it wasn't like him. He smiled, encouraging the young man to continue.

His brother spoke up instead. "I chose a different specialization, Father, at a different school: Microbiology at the University of Chicago. We finished at the same time, however, and discovered we could get onto the same faculty. So we've been living together and teaching ever since." He looked at his brother and smiled. "Not the sort of thing you'd have predicted." He took a sip of coffee.

"And where is that …" the priest asked. "… where you teach now?"

"Pittsburgh," Joe volunteered. "It's just far enough from home, but still close enough to get back easily; and it's a very good school, connected with a great medical complex. We like to say," he smiled, "that we teach lots of doctors and a few researchers. You can easily guess which we favor."

"So, what prompts your visit here in all this heat? Kinship? Or on your way somewhere?" Realizing that might sound too direct, he added: "Or, my kind of motive: Miss Annie's cooking?" He smiled to underline the modest humor.

"Well, we haven't seen our aunt in so long, and we don't have summer classes. With teaching assistants, our research is in capable enough hands, certainly for a few days. So we took advantage of Independence Day to get a nice break without losing many working days." Joe nodded in endorsement.

They're staying that *long,* the priest noted wryly. *It's more than relationship!* "I hope you brought books to read. You won't find much to do around here, I'm afraid, unless you plan to go further south into the Shawnee National Forest, or to Saint Louis. I'd be glad to suggest things to see and do in Southern Illinois. As to St. Louis, I imagine you know it well," he said, looking at Joe. "Although," he added, "it's certainly changed since you were in school."

Both brightened noticeably, but Tony spoke. "Actually, we hope to see both places. We're here through tomorrow but would like ideas for after that."

"Well, in Saint Louis, the Arch is a must, *and* the museum beneath it. The Art Museum in Forest Park is still fine; and there's the Planetarium and a new Science Center. And for flowers, besides the Jewel Box in the Park, you can't beat Shaw's Garden southeast of there."

"The Cardinals may be in town, but that's easily checked. There are *three* Cathedrals worth seeing, plus the magnificent Benedictine Priory church in the county *and* a fine synagogue near it. Even if you've seen them all, I imagine Tony hasn't. And as for restaurants, it depends upon your wallet and choice of cuisine. But there are more than enough for a couple of days."

"What about further south? Don't they call that the Illinois Ozarks?"

"Yes, they do. Giant City, the Garden of the Gods, Pine Hills, the Little Grand Canyon. If you like small towns, Elizabethtown's nice. On the Ohio, it has an interesting small hotel and some really fine, little-known historical lore, plus at least one good place to eat. Metropolis has Superman stuff and an old fort, and it's near Paducah. Cairo boasts gorgeous homes, a museum, and absolutely great southern barbecue, not to mention some of the purest southern accents outside Dixie. There's slave and Indian stuff everywhere, and lots of antique shops. And if *quaint's* to your taste, there are so many tiny places, I'd rather know which highways you want and how much time you've got before I start reciting. Although, you shouldn't overlook Anna's Liberty Pig grave," he said with a twinkle in his eye.

"The what?" Joe asked with a disbelieving look on his face.

"The Liberty Pig's buried in Anna," he said slowly, with the beginnings of a smile on his face. "It was a World War II Liberty Bond mascot and toured all over raising money for the war. It's buried in Anna!" By now was grinning ear-to-ear. "Honest-to-God! Right, Annie?" She nodded in amused agreement.

"Well, *that* may not be exactly our cup of tea. But I suppose we *can* eat our way through a scenic *or* a cultural tour," Joe quipped, looking at his brother. "Thanks. We'll get back to you before we leave."

Which sounded like the end to the conversation. He begged off and Annie regally saw him out. On his ambling return, via a circuitous route with purposeful detours to air-conditioned business district gossip shops, he had time to digest both breakfast and its conversation. *The most logical thing is her will,* he decided. *They're tacking down their place in her will. Maybe. Probably.*

CHAPTER III

Saint Helena's is a one-priest parish. Always was, even when young priests were abundant as Associates -- or *Assistants,* as they were known not that long ago. Just the place for John Wintermann twenty-eight years earlier, and now that he's considerably older, still just right. No shining light or member of any *in crowd* three decades before, in the interim he had, if anything, lessened in attractiveness on the *clerical ideals chart.* That's what local clergy call the benchmarks purportedly used in assigning parish priests.

Curious barriers seem to separate priests in the trenches from those in the Chancery, as though bureaucracy mysteriously transforms one into some foreign entity, different from, and over the years, increasingly divorced further from the other priests. An old-boy clerical network functions at every level of diocesan activity, but induction into the inner circle can strain even old friendships. Insiders use criteria that can alter even routine interactions.

Hence *the ideals chart,* which those in the trenches claim measure such importances as fund raising, obedience, loyalty (to Pope, Bishop, Chancery officials, doctrine, etc.), and especially confidentiality (guarding diocesan scuttlebutt). Each has its irreverent nickname: *the Rockefeller scale, the Mother-may-I factor, K-9 characteristics,* and *the I've-got-a-secret inventory.* As to this last criterion, diocesan secrets are, simply put, an endangered species. Confessional matters aside, other confidences are not kept, shall we say, scrupulously. To the presbyterate, this is the oddest item on an already strange list. Ultimately, clerical secrets are things told to only one priest -- at a time.

All in all, truly pastoral abilities seldom figure into assigning priests to parishes. Preaching/teaching, administration, counseling, personability, or even theological expertise don't seem to matter that much.

Not that these criteria would have located Father John elsewhere twenty-eight years previous! Never theologically impressive, his preaching's average and laced more with humor than brilliance. People skills are his forte. Honed by years of concern for his flock, they enable warm relationships with even the coldest crustaceans. His success lies largely in his genuine care for people -- that and his vulnerability endears him. Clearly lacking many professional qualities, he has never taken himself seriously. His has been thus an unthreatening and most engaging presence. At first encounter you know he likes you; and not long thereafter you realize the sentiment is mutual.

So, nearly three decades earlier, he wound up at Saint Helena's, the perfect place despite the system. Small enough to allow for knowing everyone, it's large enough for plenty of *people-patching,* as he describes his ministry of healing the wounded hearts of parishioner and pup-licker alike. Everyone needs healing, he knows. Even pup-lickers.

When he was a boy, Maple Grove Catholics in parochial school were taunted as *cat-lickers*. Their retort was *pup-lickers,* the epithet for public school students. Though decades out of vogue, the terms became deeply ingrained and very meaningful -- and his private code. As to religiously casual Catholics, he abandoned the terminology of youth and training alike. Neither *lapsed* nor *fallen-aways,* these were *strays.* These affectionate and quite personal terms are emotion-laden and too easily misconstrued, and require time-consuming, often ineffectual explanation. So they remained his thoroughly private preserve, not shared even with fellow priests.

Annie was one of his cat-lickers, but far from his favorite. He'd long since discovered that her haughty mien barely masked cruel and manipulative tendencies which she probably saw as necessary to her aristocratic demeanor. To him it bordered on the sinful, though it was probably ingrained, irreversible, and thus inculpable now. In any case, he kept her at an emotional distance and tried to overlook such things. Having to deal with her nonetheless, he resorted to several techniques, the most useful of which was mild flattery.

On a hunch, he paid a mid-morning visit to the mansion two days after his breakfast there. The large front porch was empty, the swing motionless in the already sultry morning heat. His knock at the front screen door produced a thin inquiry from the rear of the large house. Announcing himself to reassure the old lady, he was surprised to hear that the door was unlatched. He should not only enter, but come through to the summer porch.

There he found Annie, be-aproned and domestic, sans nephews as he had hoped, but conversing with the local junkman. The young men were off to the bottom of the state till Thursday or Friday, she explained. He couldn't resist a tongue-in-cheek inquiry about the safety of an unlocked front screen. *Horace came to the back door just minutes earlier and I hadn't intended to be long,* she explained, perhaps missing his tease. Certainly, ignoring it entirely.

Horace Denver had been a town fixture well before Father John's own Algoma advent. In his fifties, Horace was an interesting admixture of health and genetic misadventure. A short, thickly built man, his whole left side from the face down sagged as if some terrible weight were tugging at his ear. As a result, he walked with something of a limp, yet seemed to possess significant upper-body strength. Town children told their juniors wild tales about his occasional ferocity, but Father John treated those as local myths, barely believable and based loosely upon his physiognomy and dress. He was a man of very few words and regularly worked the town's alleys, carting off trash to salvage by trade, sale, or more recently, recycling. His base was a junkyard he apparently owned near the branch at the city limits. And while rumored to be financially comfortable, he dressed like a working class drudge, even for Sunday Mass, which he attended weekly without fail.

Father Wintermann had long since quit giving him a set of collection envelopes because he never used them. And since it was his pastoral practice never to discuss donations with parishioners, he assumed rumors of Horace's financial security were just that: rumors. He'd always found the man soft-spoken, congenial and terse -- often monosyllabic, in fact. But he never failed to attempt conversation, convinced, as he was, that it was a valued and welcome break in the man's assuredly lonely life.

Horace lived alone save for a proverbial junkyard dog. His latest in a series of large canines seemed never to leave the yard's fenced confines, and on the few occasions when Father Wintermann ventured near the premises, it eyed him without barking. Its size an obvious insurance policy, Father John kept a respectful distance, despite suspicions that the animal's ferocity were as unfounded as the rumors about its master. The only time he was ever inside the fence, the dog had -- at Horace's invitation -- quietly obeyed a hand signal and lay down nearby, after receiving a hesitant pat on the head that the priest was more than encouraged to bestow.

Horace had no other apparent companionship, since people stopped past only on business. Nor was he invited anywhere, although he did seem to spend periodic time on Annie's back porch. One fall Father John had shared late afternoon tea with him there, something Annie's demeanor at the time suggested occurred at least occasionally. He chalked it up to the complexity of the human psyche, Annie's reputation being hardly that of town humanitarian.

Well, here he was again on the fan-cooled summer porch. But no food or beverage was in evidence. Horace was seated in a solid wooden chair, conversing with Annie. The ragman -- Annie insisted on the term as less plebeian and more genteel than anything containing the word junk -- made an effort to stand as the priest stepped into sight. Father John quickly motioned for him to remain seated. "I was in the neighborhood and decided to stop. By all means, continue your conversation."

But Horace rose anyway, excused himself and was out the door in seconds, shuffling down the back walk after brushing past the priest and shouldering a bundle just outside the screen door. *Some clothes the poor man can perhaps wear,* was Annie's cryptic comment once Horace was out of earshot. From her nephews, the priest guessed, though it didn't appear much they owned could fit the larger man.

"What is on your mind this warm summer morning, Father Pastor?" Annie asked primly. "Would you like some iced tea?"

"No, thank you," he replied. "I was just wondering about your nephews. They've gone, you say. But where to, exactly? I mean, any place I mentioned? They never called. I had some other ideas..." He took out a large handkerchief to wipe his damp forehead. "The fan feels nice," he added, glancing upwards as he stood damply in his black suit directly under its stream of air.

"Yes. Went to Carbondale to visit SIU, and they intend to see some of the scenic spots you mentioned. They decided against calling, since the university folks could direct them equally well once they got down there."

So they could, he agreed. Then, feeling the heat, he announced a change of heart about tea. It would also guarantee remaining in the privacy of the summer porch, where he could leisurely learn more about the young men without alerting the neighborhood. Returning his handkerchief to his back pocket, he sat down in the sturdy chair vacated by Horace and was soon sipping in silence from a large tumbler. Annie waited for him to speak.

After an artful comment about her new permanent, Father John returned to the subject of her guests and soon had more of what he'd come for. Over a single glassful he got middle names, actual ages -- Joe was older by a year at thirty-three -- and the fact that

Tony was dating, though not seriously, in Annie's considered estimation.

"You must be very fond of them, your only nephews and all," he said. "Probably tempted to put them into your will. I'd be," he added, after a long and awkward silence on Annie's part.

"Yes, they're darlings, and, as you say, they *are* my only nephews. But I really hadn't thought of leaving them anything. Nor has the thought crossed their minds, I suspect. They certainly don't need that, single and all as they are and with such fine university positions. Aren't they nice to spend time with an elderly woman," she said, sweetly, attempting to adroitly dismiss the subject.

"Undeniably," he agreed, but not without his own suspicions, despite her claim they had no inheritance designs. *Not yet,* he thought. "Is your sister, their grandmother, still alive?"

"No. I thought I mentioned that the other day."

"Perhaps you did -- I don't recall," he said, mildly embarrassed. As she intended him to be!

"Yes, she -- and her husband also -- are gone a handful of years already. *And* the boys' parents. The two boys are all alone now. In fact, they finished college with their grandparents' assistance. My sister, Adeline, and her husband, Robert, helped them considerably, since the boys' parents couldn't. An attorney, Robert was, if you didn't know. And well enough off, I might add."

"Is that all your family, then?"

"Yes. Just my sister, myself and one other sister who never married and is gone also. I never married either, as you realize."

"Then they're your only heirs. Why not leave them the house and all?" Once again silence, which prompted a disclaimer. "Pardon me. I've overstepped myself. Thinking out loud, I'm afraid. It's none of my business."

"Well, you're certainly correct about that," she said with significant snippiness, the iron fist momentarily peeking out from beneath the velvet glove. Then in softer, more conciliatory tones, she added: "But it's not an odd thing to consider, I s'pose. But they're not in need, as I said. And they'd never move here to use the house. Understandably, the boys just don't regard this as home. Anyway, I've always thought more of *Algoma* charities. "

She had clearly closed the subject and further probing would be decidedly indelicate. He was tempted to think she was trifling with him, though her face gave no indication. He knew it wasn't beyond her to tease about Saint Helena's being among her possible *charities*. But on that subject there was to be no further elaboration, and certainly no further discussion. Not this day.

At his prompting, she indicated her parade-viewing stand on the Fourth to be her porch, as usual. *The fireworks will, of course, be much too late.* And with that, after refusing a refill, the priest stood to take his leave. *To be about parish business* -- his standard line.

On the slow, warm walk back he couldn't dismiss the thought that the lads hadn't come for professional reasons; or purely familial ones either. *If it is the will, Annie'll prove a tough nut to crack.*

CHAPTER IV

Far from all in Father Wintermann's flock were senior citizens. There were some younger families, but since the parochial school closed a decade earlier, their children attended public school. So, Saint Helena's and the neighboring parish shared a weekly parish school of religion run by Saint Edward's Director of Religious Education, a nun. The program spanned the school year, and in summer there was a week's ecumenical Bible School for tykes, plus field trips for altar servers and various student groups. On the last Tuesday of June, the high-schoolers were at Six Flags in Saint Louis County. That same hot summer day, Horace Denver was rushed to the Burger hospital.

Father John doesn't do field trips. Lay volunteers are solicited for chaperone duty. So he was at the rectory when the afternoon call came from the hospital. He was surprised to hear it concerned Horace, but didn't waste time on details. As it turned out, Horace didn't need anointing; but he did have a badly gashed right arm. A large, rusty sheet of metal had slipped and put him in need of stitches, a tetanus shot, and perhaps a place to stay for several days. As Father John arrived, the doctor was insisting that Horace rest his arm and take a cautionary dose of antibiotics.

Horace's left side, while deformed from birth, was not paralyzed. But that arm was decidedly weaker than the other. With his good arm now temporarily in a sling, he'd be hard pressed to do any serious lifting. In fact, it was questionable that he could even manage alone the next few days. Doctor Ferlin was all but insisting he remain at Saint Luke's or, failing that, with someone who could assist him and guarantee the medication regimen. Horace was having none of it as Father Wintermann entered the E. R., to anoint a dying man, for all he knew. He could see instantly that anointing wasn't needed. Physical restraint, perhaps; but not the oils! As the men argued, Father John was briefed by a nurse and then joined the discussion.

"Hello, Horace; Doctor Ferlin. I came as quickly as I could." Neither man had noticed his arrival, but it was Horace who was momentarily startled, then relieved, to see the priest. Before he could speak, however, Father John continued, addressing no one in particular: "Horace needs to stay somewhere a few days? What about with me?"

Ever monosyllabic, Horace grunted a quick yes, nodding vigorously for emphasis. Anything to get away from the hospital, it seemed.

The doctor took some time to reply. "Fine with me, Father. *So long as* you provide the required therapy," he said with a determined look on his face.

"Which would be ... "

"No lifting with the injured arm till I say so; rest today and tomorrow; and sticking precisely to the medication I'm ordering. I've given Horace a tetanus shot, but I'm also putting him on precautionary antibiotics, which are to be taken as prescribed and the dosage completed. Can you guarantee Horace's adherence to all this till I remove his stitches?"

"I don't see why not. Wouldn't you agree, Horace?" The injured man gave another exaggerated affirmative nod and grunt, too hastily in the priest's opinion. But even if Horace was promising the moon to be rid of things medical, the priest felt he could handle him.

The doctor hesitated, perhaps from similar suspicions about Horace's sincerity, then agreed he could accompany the priest -- after exacting a formal promise from Horace to visit his office in four days *and* as needed thereafter, till he could certify him healed. Under the circumstances, what else could he do!

Horace was silent on the return trip past corn and bean fields stitched alongside the highway, heat waves undulating unevenly above each design. *Crops, at least, thrive in this heat!* In the silence, the priest reflected on the luxury of air-conditioning in car *and* home, and wondered fleetingly how his family had ever managed without it while he was growing up.

The priest didn't press the junkman till they neared the edge of town. Then he asked: "What about your dog?"

"I locked up before going to Burger. There's food and water. He'll stay on guard. We can check tomorrow." ... *on both dog and yard* the priest thought, even as he wondered about the *we* part. *Cross that bridge tomorrow -- for tonight, supper!*

A man of simple tastes, Horace was easily pleased. They'd dine in on funeral hot dogs, a peculiarly local delicacy so named because they were served after funerals by area parishes, and were generally available at retail outlets. *A lot like Ball Park Franks,* Father Wintermann loved to explain to quizzical outsiders whenever they'd question signs outside grocery stores -- he pronounced it like the local he was: *groshery*. Supermarkets had yet to invade the county, though one was rumored for Algoma. He'd believe it when he saw it!

During the next few hours Horace proved surprisingly docile, and with minimal post supper conversation over television, he made for bed at eight o'clock. After overseeing pill taking and determining Horace's expected hour for rising, the priest bade him goodnight. By ten o'clock when he himself turned in, he had evolved several Horace strategies for the morrow.

Horace's day followed the sun, so Father John was up earlier than usual and had time to bribe his patient into attending Mass with the promise of breakfast at Annie's. She hadn't heard about Horace's accident till the priest's pre-Mass phone call, so breakfast was accompanied by solicitous inquiries about the incident, Horace's current status, and the prognosis -- altogether more concern than expected or warranted, in the priest's opinion. Nor did that cease at meal's end. Since Horace's medication was in his pocket, Annie's surprise invitation to remain not only could be accepted, but had to be. After, that is, promises to stay put, avoid lifting and continue the pills.

Back at the rectory, the priest didn't know what to make of the intensity of Annie's interest and was still puzzled as he reclaimed Horace in late afternoon. Annie had apparently managed to rein in the man, which provided even further amazement. Nor did Horace offer any helpful insights into those minor mysteries as he and the priest checked on dog and yard. Horace posted a crude sign announcing the yard's temporary closure, the absence of same

29

seeming to have posed no problem the intervening day. Before bedtime, Father John was able to elicit more conversation, if only a bit more.

"What's your dog's name, Horace?" he asked during a break in the very old Western Horace had selected. Father John rarely used his TV and readily allowed visitors to decide any viewing fare.

Without looking away from the TV set, he said: "Dog."

Why is that no surprise? "And how long have you had him?" the priest asked with a straight face: no small feat!

"He's ten. Had him since he was a pup." Then after a moment's pause, he added: "Miss Annie helped me get him. She said I needed a really big dog so people would think it was mean."

Annie again!

Knowing full well she didn't especially like animals, the priest prodded, again with an innocent face: "I didn't know Miss Annie ever raised dogs."

"Oh, she don't; she only helped me find it. A friend had a litter, she said." He paused, then added quietly: "I really think she bought it. Over in Burger."

"That was very nice. Miss Annie seems to do a lot of nice things for you, Horace. She must really like you."

"When I came here they told me to see her, cause she might help get me a job. She did, sure enough. Took me to the junkyard and Mr. Simpson hired me right off."

"When did old Mister Simpson die? Before I came, I think." He tried to sound nonchalant.

"Yep. Think so. Anyway, it's when I got the yard. When he died."

"So, where'd you come from then, Horace?" The priest's feigned ignorance went apparently unnoticed, the basics of Horace's story being common knowledge in town.

"Colorado."

"Got relatives back there, do you?"

"Nope. I was a orphan." He finally looked over at the priest.

"I'm sorry to hear that. Did you know your parents at all?"

"Nah. All I know is I growed up in Saint Gertrude's Home. They sent me here when I got too old to stay."

"Twenty one?"

"Think so." And after a pause: "I liked it there; they was nice."

"Saint Gertrude's? In Colorado Springs, wasn't it, Horace?"

"I guess. The mountains sure was pretty. And the Sisters was real nice to me. Let me work on the lawns." Again he paused. "The kids wasn't always nice, though."

"I think it's closed now, isn't it?"

"Don't know." Horace said, as he shifted a small pillow behind his back. "Ain't been back," he said matter-of-factly.

"Didn't you ever talk to anyone there after you left -- on the phone, I mean? Or maybe write your friends?" He tried not to sound accusing. But upon reflection, he realized Horace was hardly the letter-writing type.

"Nope." He said it as though it hadn't occurred to him or, indeed, even seemed important -- ever.

"So, why'd they send you to Illinois? We're nowhere near Colorado, are we?" He was beginning to sound doggedly determined.

That went unanswered. The program had resumed and Horace wasn't about to miss any of it. But an offer of popcorn guaranteed they'd see the whole movie, postponing Horace's bedtime and prolonging the conversation, which began again at the next commercial.

"Didn't you look for a job outside Saint Gertrude's, Horace?" the priest asked as he passed the popcorn. "I mean, when you were getting too old to stay on there?"

"Uh-uh. I was gonna work *there*. That's what the Sisters said."

"At Saint Gertrude's? You were lucky then, not to get that job, since it closed down. But how'd you happen to come here?"

"Don't know. When it came time, I just came here on the train."

"Wasn't that upsetting? That sounds pretty sudden to me."

"Nope. They said I'd like it; and I did." Horace's face exhibited no change of expression, a normal response pattern that made him well nigh undecipherable.

"What was so nice about coming here to Algoma?" The priest believed anyone else would have seen through him by now, and he worried that at any moment even Horace might.

"Mr. Simpson was nice. He gave me a dog. A different dog. This is my fourth one. And he showed me a lot about running the yard."

"Didn't he die soon after you came?"

"Nah. It was a couple years or so."

"That's kind of what I meant, Horace. Because if he died that soon, how'd you get the yard? I mean, you were still pretty young, and all. Did he give it to you in his will, maybe?"

"I guess. All I know is some lawyer told me it was mine, and I should run it the best way I knew how. Told him I would."

"You must have signed some papers?"

"Nope."

"Not *any?* "

"Maybe. Can't member, though."

The priest was disinclined to believe that, but he didn't know how to argue it further. So he changed topics. "You said Miss Annie was nice to you. From the very start, I imagine you mean?"

"Uh-huh," he agreed.

"Like how? I mean, how did you meet her?"

When Horace went silent again for the story's final segment, the priest figured he'd gotten as far as he dared. He wasn't any closer to figuring out the Annie-Horace connection, but he now realized it was long-standing. How did Saint Gertrude's know about her -- or vice versa? Was the orphanage one of Annie's charities? Annie's reputation around town wasn't one of generosity. It made no sense. There was still a lot more work on what the priest was beginning to believe was as close as he'd ever got to a real Algoma mystery.

CHAPTER V

Within three days, not four, Horace healed up well enough to satisfy the doctor and was back at the yard, after earnestly promising to finish the antibiotics and get his stitches removed. Thus was Algoma normality largely restored. Summer's oven roared on with new ways daily to bake the countryside. Annie held court on her porch some evenings. Horace scavenged the town alleys. And Father John continued to wonder about the unlikely pairing of the ragman and the duchess.

The return of the nephews the day Horace went back to work provided a fresh chance to further unravel their piece of the knot that preoccupied the priest. Just in from Saint Louis, their second area of exploration and one that kept them away longer than originally intended, they spoke to Father Wintermann at Mass the morning after their reappearance. *It was nice to visit Washington University; we spent a bit of extra time there.* Anthony volunteered that, and it struck the priest as unusual, since, as he remembered it, Joseph was the Wash U alumnus. *Perhaps it's just his enjoyment at nosing around a fresh research facility!* Father John asked if they'd be staying for the parade and fireworks.

"Unfortunately, no. We must get back to our research," Anthony said, his words conveying more regret than his voice. "We'll be spending our several remaining days with our aunt. Who knows if we'll get in another visit! She's not getting any younger, you know."

"Sounds like you think there's more to worry about than just her age," the priest joked.

"Well, she *is* having bronchial problems, in case you haven't noticed."

He hadn't.

"Isn't it darling the way Maisie refers to it: *her bronichal tubes be actin up onced a while,*" he said to his brother, who smiled. He turned back to the priest and added: "It may not be serious, but at her age and in this heat and humidity, you never know."

Maisie Brown was a young black woman who cleaned several of the town's large homes. She was a reluctantly hired intrusion at Annie's every other week. And that, only within the last year or so.

"Is your aunt doing anything about that condition?"

"Not yet, but we're trying to persuade her," Joe said, sounding determined, but unhopeful.

"Hope you succeed," the priest replied. "I don't want to lose my breakfast connection," he said with a smile. But he wondered how significant this malady was, and precisely how it should be treated. "Let me know how your campaign goes."

They promised, of course, though not in very convincing fashion.

True to their word, the nephews stayed just two more days, planning to leave immediately after Mass the third morning — *to miss the holiday traffic!* The day before they were to leave, they extended an invitation to breakfast the next morning. So it was that the trio was slowly walking back to Annie's in the already sweltering heat that last Monday morning when Father Wintermann, whose summer thoughts were often on the weather, mentioned how nice it would be if the mansion were air-conditioned.

The younger men agreed, *particularly in light of our aunt's condition.* "We tried to convince her of its importance, not just its convenience. But she won't hear of it. Couldn't even buy her a small window unit. Or a ceiling fan! So we simply put our feet down and bought a small oscillating fan for her bedside. God only knows if she'll use it." Joseph didn't look hopeful.

"You've mentioned breathing difficulties before. I assume you believe her problems are significant?" the priest asked.

"We think so," said Anthony. "I mean, with the elderly, anything like that's likely to be, don't you think?" He looked at the priest rather intently, obviously expecting agreement.

He didn't agree. But he decided against communicating that, reserving judgment to learn more than he felt he was getting from the young men. Instead, he tried to sound pastorally concerned, and clucked reassuringly as they stepped onto Annie's front porch, a non-answer answer.

Without waiting for his aunt, Joseph opened the screen door for Anthony, who announced their presence as he entered the front room. The three walked on through to the kitchen where they found Annie removing biscuits from her oven. The kitchen commander-in-chief handed them to Anthony and ordered them all to be seated on the summer porch while she got the coffee. As they stepped onto the porch, they could see everything else was at hand, an enormous meal. *A traveling meal,* she called it: a steaming platter of fried potatoes, scrambled eggs and bacon, plus toast with homemade strawberry preserves and butter alongside, the mélange crowding the modest summer porch table. As she poured coffee, Annie asked the pastor to offer a *special prayer: not only of thanks for the food, but for my nephews, and especially their safe return.* Which he did before they sat down to the farewell repast.

Father John kept a close, if guarded, eye on Annie while they ate. Her health was not mentioned, but try as he might, he saw nothing even remotely problematic. The conversation was mostly small talk about routes to follow, the time it should take, and the like. The boys promised to phone from home, and then to stay in regular contact, encouraging Annie to use the mails, *which are cheaper, auntie.*

His inability to verify the condition the nephews had so confidently mentioned initially puzzled him and then began vaguely troubling him. *What's escaping me? Real problems, however peculiar? Delusion on their parts? What?* He almost missed a brief aside to Annie as they were rising from table about using things -- *plural* -- they'd left upstairs. He was aware of only a fan. What else could there be?

After the three had helped Annie clear the table, over her mild objections -- *I can warsh the dishes myself, thank you* -- they all made their way rather quickly to the front of the house, and the young men, with a flourish of hugs and kisses, bade good-bye on the front porch. They walked out to their already packed car waiting at the curb, and in moments were gone, waving and tooting their horn down seventh street for a block or more before disappearing directly into the sun's scorching fireball above the eastern horizon.

Father John insisted on helping finish the dishes. He knew she had no automatic dishwasher -- *on principle,* as she often put it. "The least I can do, especially in light of my many meals here," he said, "is to help wash dishes." He deliberately toned down his pronunciation of the verb, different as it was from Annie's and most Algomans.

Years of seminary speech training had expunged that eccentricity, ingrained from youth as it was in many Southern Illinoisans. *Warsh the dishes in the zinc* was one of the phrases classmates had found funny. That, and words with the long "o" substitute, such as *form* and *mortar* instead of *farm* and *martyr.* And also the Germanisms of his grandparents' and even his parents' eras: *throw the horse over the fence some hay,* and *it stood in the paper.* They were all flitting through his mind as he stood by her *zinc* and helped Annie *warsh* and dry the morning's dishes.

He'd hoped to further research the conundrum during his kitchen chores, but found himself no closer when the last dish was safely put away. He reluctantly took his leave out into the blazing sun in search of cooler venues, vowing to somehow unravel this tangle.

As he walked back, he noticed small plants in yards along the way, their leaves seeming to cringe and turn in on themselves as if to escape the omnipresent heat -- *or were they just withering?* And *he* began panting in shorter and shorter breaths, which he attributed to the combined forces of heat, humidity, exertion and weight -- the weight of both frame and years. Yet, Annie, far older, had no such difficulties. And what with hot work in a hot kitchen, she *ought to* ... *if* the nephews were even remotely accurate!

It clearly didn't add up, and he redoubled his intention to get to the bottom of it, though he hadn't the remotest idea how. The several messages waiting at his rectory, however, sufficiently distracted him from further thoughts of l'affaire Verden for the rest of that day.

CHAPTER VI

Why Father John hadn't thought of Maisie before, he couldn't immediately say. When her name did pop into his head, the idea seemed so right, that he berated himself for not coming to it sooner and letting several days slip past instead. He'd been thinking more about Horace in the interim. *Which is probably why I overlooked Maisie.* In any event, the actual moment was serendipitous.

He had decided on some research at the town's weekly newspaper -- from the years before his own arrival -- to find out more about the junkyard and how Horace came to own it. On his way to the newspaper office Maisie came to mind. He was singing to himself -- something he couldn't recall later, but an old tune -- when a child rode past on his bike. He switched instantly to *A Bicycle Built for Two,* and when he got to *Daisy, Daisy,* he found *Maisie, Maisie* coming out instead. He laughed at that, and then stopped stock still, realizing he'd stumbled on something quite amazing. Well, modestly clever, anyway. He tucked the idea away and, pleased with himself, resumed his stroll rather more briskly, considering the heat, down the town's main street toward the plate glass window of the storefront proudly bannering *The Algoma Smile.*

Other priests liked to tease about the newspaper's name and especially about its propensity for providing details in the absence of elusive, dull or, especially, non-existent originals. It mattered little that small town Illinois boasts many such papers with exotic names and reportage equally as enlightened as the Smile's: weeklies with names as pretentious as *The Weekly Post Gazette and Intelligencer* or as cozy as *The New Berlin Bee.* They are all over-heavy with local business ads plus a plethora of personals, many consisting of sentimental doggerel about relatives a month, year, or even five or more years departed. They also contain abundant reports about who had tea with whom on the occasion of whatever semi-momentous occasion, particular care being given to the precise mention and spelling of everyone's name. It's the small town answer to Hedda Hopper, but with information supplied by the principals, not some snoopy outsider, so as to insure a patina of prestige, however self-manufactured, and also to tweak the pompous or eccentric as part of a region-wide game of self-aggrandizement through diminishment of any-and-everyone else: details simply too important to leave to reporters.

As he entered the newspaper office he was greeted with *How's Father Wintermann today?* The third person vocative never failed to bring an interior wince. As normal in Algoma as it was unusual in Maple Grove, that speech pattern, like many other things in Southern Illinois, was probably Appalachian in origin, the product of migration from western Kentucky, often as not via southern Indiana into southeastern Illinois and onward. In search of livelihood, the poor brought along customs, speech habits, even recipes. By now the fallout provided a mélange which included country-western music, pick-up trucks, bow hunting for deer, a distinct spoken twang, the occasional cowboy hat and boots, ham-beans-and-cornbread -- and third person greetings.

He was forever tempted to respond in the third person, in hopes the irony wouldn't be lost; but it was de rigueur not to. Rules of engagement allowed only the initiator to employ what was intended to be creative and endearing. Replies had to be in the first person. Craftily he said, "Not so well today, Herb, in all this heat," in an effort to test another of his pet peeves: people not wanting literal answers to such questions. Herb's surprised look of combined annoyance and empathy convinced the priest he'd got it right.

"No health problems, I hope," the editor asked with some concern.

"No, not that. Just wish I could get the Almighty to air-condition the outdoors as well as you've done the inside of your office. I doubt if rotten apples could even breed maggots out there; not without the little critters getting fried." He smiled, taking Herb rather easily off the hook he'd all too briefly had him on.

"Oh, we get flies in this heat. They're just growed up maggots. So I'm bettin we *could* get maggots on an apple. *But*, if you let it on the sidewalk for, say, half a day, maybe you're right: apple and maggots both would fry right up!" He smiled an editorial smile and made a mental note to save the thought for some future edition. "What brings you here? Besides my air-conditioning."

"I'm hoping you've got copies of the paper from perhaps thirty years ago. What do they call it at big city newspapers? A morgue?"

"You're in luck. We sure do. But you gotta keep 'em in the building, *and especially* you gotta go through 'em slow and careful.

Them puppies are gettin right brittle. What the heck you interested in thirty years ago for?"

"Just reckon it's time to find out more about what produced the parish I inherited, Herb. You know, sort of professional curiosity." As someone who thought *new* meant anything since the Bicentennial, he hoped the editor didn't think thirty years ago old *enough* to question him further. He also hoped he sounded plausible enough, or that Herb was busy enough to let him be.

He had. Or Herb was. In any case, the files were shown without further complication or comment, and he settled in for a slow dredging through the *Archives,* as the label proudly identified each sheaf of issues. He began with the year he arrived, intending to move back year by year as far as necessary. It might cover as many as ten or twelve years. He couldn't be certain.

What he began to find was interesting, even fascinating, but not what he came for. He resigned himself to spending a long while at his task -- perhaps several long whiles -- as he kept being diverted by trivia. *Algoma Patchworkers entertained the Ladies' Quilting Circles of Frogtown and Beaver Prairie at the Verden Mansion last Tuesday. Before the delicious lunch served by Miss Annie Verden, many examples of the quilting art were displayed and inventive new ideas were exchanged. In all, twenty-three ladies were present ...*

Annie was mentioned regularly, usually entertaining someone at her mansion, with the one exception of an item detailing the exquisite care for her flowers, a hobby much neglected now, the priest noted. But Horace, the Simpsons, and the junkyard were nowhere to be found in that year's newsprint.

43

The previous year brought to light nothing of substance either, though familiar names continued to surface, involved in things like hunting trips to the Cache River Scatters near Vienna; or land sales, especially one featuring a Missouri businessman who, it could be surmised, robbed someone of a birthright, nefariously removing land from rightful control of the family that had it forever: a rather early chapter, it seems, in the demise of the family farm.

There were death notices too, and he suddenly realized he should have been searching them for old man Simpson. He went back through both years with that in mind, but without success: still the Simpson name failed to appear.

He looked at his watch and realized he was dangerously near missing an appointment. Thanking Herb, and promising to return soon, he abruptly made for home, walking faster than usual, given the sting of the late morning sun. He arrived perspiring and out of breath to find an understanding Father Harold Fick nonchalantly waiting in a lawn chair on the shaded rectory porch.

Middle-aged and younger than Father John, Harold fancied himself a clerical Renaissance figure surrounded by a Philistine presbyterate. He was in fact something rather more rococo amidst a bevy of clerical Calvinists who lacked the faintest sense of humor in his regard. The more seriously he took himself, the more they stared in disapproval -- incomprehension, as he saw it. This much was so: Harold stood out, though not so much for praise *or* blame! He was just hilariously out of synch. A good snicker or uncontrolled guffaw would have served better than the slacked jaws and quiet stares he usually evoked and always misconstrued. If not his attire, it was his conversational material, or his gangly, disjointed gait or ... well, you surely get the idea!

He was genuinely nice. But think of the mileage to be gotten from realizing the truth and playing it for its humor. Instead, he plodded on, hindered by the ignorant, as he saw it, while in fact being disdained by indifferent souls who felt superior. It was a tossup who was the more misguided.

John Wintermann could never observe this without a twinge of sadness and pity, but mostly he felt ingratiated into the man's earnestness, and he sincerely accepted and befriended him. He was the one priest with whom Harold felt truly comfortable. While most saw him as good for a surreptitious laugh at best, too kind for that, John Wintermann laughed -- if at all -- only with Harold. Had they been stationed nearer and able to enjoy each other's company oftener, they could have been the diocesan odd couple. As it was, John's simple, quiet friendship was welcomed though not completely understood. He couldn't bring himself to point out Harold's many incongruities. He simply paid him the ultimate compliment of accepting him for himself.

The day was to be dedicated to thorough-going self-indulgence: the Saint Louis art museum, an early evening Italian meal on the Hill and, despite the heat, some Broadway production or other at the Muny -- John couldn't remember what; but Harold, as always, knew. All of that, *after* the blue plate lunch at the Algoma truck stop. And, for a day, Father John put Algoma and Saint Helena's completely aside, including Annie, Horace, and Ichabod Simpson.

CHAPTER VII

Several days slipped past before Father John could return to his self-imposed sleuthing in the Smile's morgue. A funeral had intervened, as well as the summer Clergy Conference concerning diocesan finances. The funeral was the more humane of the two. On the other hand, the clergy gathering was air-conditioned.

On the way to the cemetery, the funeral director's large automobile contained just two people. Father Wintermann had long since discontinued dragging servers to burials, so his sole companion that day was his parishioner, Larry Feldspar. The senior owner of the funeral home was older by far than the priest and heavier, as well. As he watched the hearse ahead of him being driven by his son and now principal operator of the mortuary, he also kept a watchful eye on the long cavalcade behind him. He had the leisure on that slow, reverential drive through town and onto the desolate cemetery lane to indulge his propensity for reliving the good old days for anyone who would listen -- in this case, his pastor. "Did I ever tell you about one funeral I attended as a boy when my father was running the business?" he asked.

"Which one would that be?" Father John responded, in his best straight man mode. Larry was a large man with bushy eyebrows and a nose that suggested his girth came from more than rich food. Father John knew from long experience that Lawrence M, as the nameplate on the firm's vehicle identified him, was a man who enjoyed life -- every bit of it, including its humor.

"Well, they didn't have artificial grass, ek-cetra, around open graves those days," he began, "and there wasn't no apparatus for the casket either. Caskets were on planks or ropes over the opening -- we preferred planks *and* ropes, cause if you just used ropes, a couple big fellows had to hold 'em during the whole graveside service which might get kinda long. And with our weather, it wasn't altogether satisfactory -- slippery sometimes, you know -- so we used both planks *and* ropes. Anyway," he paused for a deep breath, "ever'body was standing round the grave. There wasn't no tents then either, unless the deceased was well off, which wasn't very often, a'course. So, ever'body was there in the open, and just as the minister -- pretty sure it was a Protestant burial -- just as he began, an old man sneezed, and his uppers popped outen his mouth and into the open grave." He began laughing at the memory, and it was moments before he could continue.

"Well, the minister didn't zactly see it -- though most of us others did -- he just heard the sneeze -- and he began his prayin. The old man was embarrassed no end, don't ya know, but he was also upset about the teeth. And, you might guess, the teeth won out over his embarrassment. He didn't know what to do at first, how to put it to the minister or get his attention. So he used his cane to pound on the casket, which, let me tell you, certainly got *ever'body's* attention."

"Now I's just a young pup and was fit to be tied. My momma was poundin on my head to stop me from snickerin, though she could hardly keep from it herself, when the minister draws himself up to full stature and asks what the man thinks he's doing disruptin such a revrent moment. The old guy brings the house down by lisping that he *paid a good thirty-two dollar for that plate and darn well wouldn't donate it to Hiram's cause, no matter how grieved he was over his passin.* Even the widow giggled. Course, they had to remove the casket from the planks, and then a man lowered me into the grave so's I could fetch the teeth, though I remember feelin all icky about 'em, full of spit and dirt like they was. The minister shortened the service once we all got our decorum back, and my father later joked about maybe payin the old man to do that at every Protestant burial just so's to hurry 'em up." By this time he was guffawing out of control, and Father John thought he'd maybe have to grab the wheel to keep them out of the substantial drainage ditch just before the cemetery arch.

Moments later they pulled to a stop behind the hearse and were waiting for everyone to get out of their cars. Father John squinted into the light blue haze created by the sun beating on the already hot earth and glinting blindingly off the even hotter marble headstones, and remarked as casually as he could: "I can take a hint: I'll keep the service short," which, because of the weather, he had already intended anyway.

The older man stifled another laugh but an instant later opened his car door with the sort of deadpan face Buster Keaton might envy, an ability, he later claimed, that was a professional secret. But the performance so tickled Father John that he remained in the car himself a moment or two to regain his own composure. The committal service was indeed brief, but the priest's compassion was as genuine as always.

The Clergy Conference was another matter. For as long back as he could remember, Father Wintermann had always felt ambivalent about such gatherings. He enjoyed sharing gossip and renewing acquaintances, but felt only the occasional professional topic worth his time or anyone else's. He particularly disliked hearing about diocesan finances, but not for lack of understanding *or* concern. Rather he thought there wasn't much any of them could really do about such things, as good or bad as any year's situation might be. It was already a fait accompli. What was spent was gone; what would be spent was already determined -- by bishop or one diocesan board or other, people he generally trusted anyway. So there was no good reason to waste half a day on such matters. It could as easily be printed for them all to read. *At least that's how I see it* he'd say to the rare someone who might ask his opinion.

But he was of a generation that saw attendance at such things as a matter of common courtesy, if not obligation. And though younger men were more and more ducking such moments, he joined a number of other priests in the DuQuoin parish hall and put on his best face to mask his boredom. At the light lunch afterwards, the SIU Newman chaplain made small talk with him about two visiting academics he'd seen in Carbondale the week before. *They said they met you in Algoma.*

"Yes, visiting their aunt -- great aunt, actually. They just went back East."

Ed Wallen was a bright and energetic young priest and was immensely popular. "They *said* they wouldn't be here long," he continued. "A Chem prof told me they really bent his mind about the properties of gas."

"What would make gasoline so interesting?"

"No, gas -- as in chemical gases. They must've rung the changes on the subject: pressurization, safety -- even portability! Of all kinds of gasses, I guess. He really enjoyed the conversation because it was so different. Not the usual sort of thing they talk about, I suppose. Genuinely interesting guys, according to him."

"How'd you meet them, Ed?"

"They came to a weekday Mass."

"They did in Algoma, too. Didn't talk much science with them, though."

"I didn't either. That was the chemistry prof. They just asked about Pomona and the covered bridge. And whether there were any festivals nearby -- you know, like Ridgway's Popcorn Festival."

"They knew about Ridgway and the Popcorn Festival?"

"No, I just threw that in now. Anyway, I said there weren't any just now. But I did tell them how to get to Pomona, and I think they were going on to Little Grand Canyon too. Oh, yeah, and I remember they asked if I'd ever been in a tornado! Had a good laugh about that! I told 'em they didn't really wanna do that. Then I told 'em what it was like: the yellowish sky beforehand; the pukey green color when it hits; and the weird, swirling winds even at the twister's edges. We talked a good twenty minutes. But no science." Just then another priest came up and pulled the younger man away, and Father John was left to contemplate the nephews' interest in gases and their containment. It seemed quirky but nothing more.

So, after a two-day break from his research project at the Smile, Father John was thinking of returning there when he remembered Maisie. It took several phone calls to finally learn that Maisie was cleaning a large home in town. He asked that she return his call at her convenience. Then, because his mind worked that way -- that is, with that ball now in another court and his presumably not having to attend to it anymore -- he promptly put her out of mind. Again. It would be another week before she'd be back in his thoughts.

With the Maisie matter well enough in hand, he set his mind once again on the Smile Archives, only to be sidetracked by a flurry of pastoral things requiring more or less immediate attention, including a long over-due annulment that one of the principals just phoned to say was finally her top priority; and some sick calls -- mostly from anxious relatives wanting Eucharist and pastoral consolation for aged relatives; and a request from Annie to *please come discuss something of small consequence.* He was doubtful about how consequential it might turn out to be, and decided to put it last on his list, in case it required more time or effort than she'd implied. It turned out to be curious but, as she had said, of minimal consequence. However, it involved Horace.

Annie hadn't seen the man in some days and was worried that his mishap had been more serious than generally thought. *Would you check discreetly, Father John -- without embarrassing the poor man by bringing me into it? I mean, he may be all right, after all* There it was again: her Horace fixation.

51

What could he do but agree. So, the very moment he left Annie, he went in search of the junkman. With a what-the-heck-why-not attitude, he decided to kill two birds with one stone and pump Horace again while he was at it.

But he didn't find him. Dog was there, properly on guard, if you can use the phrase. He wouldn't stir from the shade of the shed Horace called his office, though Father John didn't tempt him by coming anywhere near the locked gate, either. Several shouts failed to rouse Horace, and the gate lock firmly in place during business hours confirmed his absence. He decided against leaving a note, as Horace might not phone back anyway. And with his absence auguring at least good enough health, the priest let it go at that for the time being. But to be absolutely certain, he'd have to return. That's all there was to it, what with Annie involved!

He did so twice the next day before finding Horace, each time passing the big red Mobil horse still flying on the ancient glass-topped pump at the abandoned gas station just inside city limits. The temperature and the fourteen blocks to the yard had convinced him to go by car, something he ordinarily didn't do inside the confines of Algoma, preferring the exercise and personal contact walking provides. After his first fruitless trip through the hot streets, he mentally congratulated himself on that decision.

On his second effort, however, he realized belatedly that he didn't know how to begin the conversation. He slowed his amble into the junkyard, preoccupied with preparing something plausible, and almost walked right into Dog. Startled at the sight of the large animal lying directly in front of him, he paused, and received only a languid blink from the sentinel, who continued panting in one of his usual shady spots. The priest dropped back two steps anyway and shouted for Horace. The reply emanated not from the office, but from somewhere near the rear of the yard, within a hallowed recess of scrap metal piled overly high, the very place he had injured the arm Father John was checking on. The priest went round the shed and saw the junkman coming toward him, his face graced by the faintest grin of greeting.

"'Lo, Father," was all he said, but the smile perdured, small, yet pleasant and genuine. For all the heat, Horace had on a long-sleeved flannel shirt and his ever-present seed corn cap.

"Hello, Horace. Not here on business, so don't hurry in all this heat," he began, momentarily abandoning his hastily prepared script. Then returning to it, he said: "Just stopped by to see how you're doing and if the stitches are out. You look fit enough from here."

Horace raised his shirtsleeve in silent reply, the evidence plain enough: neat, clear scars: no black threads, and no discoloration.

"Wonderful. So, the arm feels all right now?"

Yep was all the priest got from him.

"I suppose, then, you're back on your rounds. Just haven't seen you the past few days," he said truthfully enough, remembering to forego mentioning Annie. "And I didn't know if I needed to buy more popcorn." Horace, appearing to get neither the allusion nor its humor, didn't laugh and continued standing there sporting the same slight smile, which was beginning to look silly.

"Well then," the priest said nervously, "I guess I can be off. Just wanted to be sure you're okay," he said, turning to backtrack toward the gate.

Before he could complete the maneuver, however, Horace said: "Got sump'n for you, Father," and stepping toward the priest, he moved beyond him into the shed. He was in and out so quickly that what he emerged with must have been immediately inside, waiting perhaps for just such an opportunity. It was a wall hanging, and Horace offered it to the priest with modest pride. "Just found this," he said, as though he had meant to add *on my rounds*. There was no trace of embarrassment, for all that. "It's in good condition and I want you to have it for bein so nice last week."

It was the Last Supper in an elaborate gilded frame and painted in a light-filled, unconventional manner. Touched, the priest debated how to respond, then said simply: "Why, thank you, Horace. You needn't have gone to the trouble ... "

"But I wanted to, Father."

Of course he did, you ninny. Just accept the gift, John! Then, aloud, he added: "Well, thanks again, Horace. I'm moved; and I'll treat it specially. In fact, I know just where I'll hang it. When you're near the rectory, ring my bell and I'll show you." Then he turned and walked out to his black, five-year-old Ford.

At the rectory he replaced the picture in the small entrance lobby. *Actually it looks rather nice there,* he thought. *I wonder if I'll have to mention it, or if people will notice without that?* He had to call attention to it, as it turned out. And in due time he showed it to Horace, who repeated his shy grin, obviously gratified he'd been able to give something special to his pastor. Over the next few weeks, Father John continued to show it off, always claiming it as Horace's gift. It was to be yet another several weeks, however, before he'd look at it closely again himself.

The Horace issue was laid to rest after he phoned Annie to reassure her of the man's good health. Her curious response was that she hoped he'd stop past soon -- to pick up a few things. The priest assured her there was no reason he shouldn't be doing that soon.

A thunderstorm popped up -- suddenly, as happens in the land between the two great rivers -- so he spent the rest of the day puttering around the rectory, putting off a visit to the Smile until the morrow. Maisie still hadn't called, but neither had she popped back into his mind.

CHAPTER VIII

Despite the rain, next day's temperature and humidity were still at record highs. Thunderstorms in the low-lying Mississippi and Ohio Valleys rarely bring weather shifts; only massive frontal movements do that. And neither the jet stream nor any serious fronts had affected their world for weeks.

With the early morning heat, Mass attendance was sparse, even though the weekend previous Father John had suggested more prayer for divine relief. He had actually said *inhuman weather,* and worried later it was too provocative -- not to his people, but to the Lord. Goodness knows, the people agreed wholeheartedly. Even farmers were sick and tired of the heat. And with occasional rains -- yesterday's was a bit heavy for purists -- the crops continued looking not only good, but perhaps record-setting. No, by now all his people felt comfortable praying for cooler weather. It was God he worried about.

After a light breakfast he steeled himself for the walk uptown and set out toward the newspaper to cool off and continue his systematic search. Once again colorful diversions flooded forth from the volume in his hands. *Eldon Blackwood caught a sixteen-pound channel cat and a mess of crawdads in the Bar Pit south of Algoma last Saturday. They were served Cajun-style Sunday afternoon to his family and neighbors by his cousins visiting from Louisiana ...*

He daydreamed, recalling what an elderly George Haderlein had told him about why those ponds are called *bar pits*. He remembered the moment well, the widower's small front porch of a summer evening, the iced tea and the reminiscences about long before the man was a State Fire Marshall, when he was *still just a kid -- back in nineteen and eleven or twelve.* Working with *a bunch of Irish and Eyetalians,* he helped construc the Beckemeyer-Breese levee road which became Route #50, their local piece of the transcontinental federal highway linking them to Cincinnati, Saint Louis, and points beyond to both coasts. He'd received fifty cents a day as water boy for the crew and was proud of the job because *those days that was good money.*

Father John knew the road now as a lovely, tree shaded mile or so drive over the Shoal creek bottoms. The only other one remotely like it that he knew of was State Route #177 east of Mascoutah. But it didn't form a restful, shaded canopy. George explained how they hauled the dirt for the roadbed up from below and then planted trees alongside the new road to hold the steep slopes in place. Soon the trees formed a lovely tunnel. *Engineers don't do that any more. Too unsafe! There'll be hell to pay when those trees come down.* The dirt, he said, was brought up in wheelbarrows, and the pools formed below were *barrow pits.* That soon became *bear* pits and in some places later *bar* pits.

But the area below that particular road *isn't what you'd call a proper bar pit,* he explained. It's all bottomland and only wet when the creek floods. The roads south of Algoma sport pools of normally very brackish water most of the year: real mosquito breeders. *They* could rightly be called *bear pits.* But those were already there before his time, George had recalled. And the only thing *he* ever heard them called was *bear pits*. He couldn't rightly say when that changed to *bar* pits, but he thought it might have something to do with pioneer slang for a bear; or maybe it just was some other kind of sound shift.

"I guess," he went on, "it's as mysterious as callin the creek in town a *branch*. It's really a creek; but ever since I's a kid no one ever called it anythin but *the branch*. I s'pose it could really be a branch of some creek or river upstream. Don't know. Never checked into it. Only sense I ever made of it was that branches were maybe shorter versions of creeks. Or smaller, narrower."

The priest went back to his reading. *The peonies are particularly beautiful this summer, and none are lovelier or larger than those gracing the garden of Florabelle Williams. Her husband, Goose, who prefers calling them* pinies, *claims their quality is due to farm animal manure which, he also claims, helps control the ant population at their roots ... Simp Simpson was taken to the Burger hospital for a wound to his lowest abdomen. While working under his car, he dropped his electric drill -- which was still engaged -- onto himself ...* Father John blushed in an unsuccessful attempt to keep from laughing at the paper's lengthy depiction which left little to the imagination as to exactly which body part had sustained the amateur mechanic's unfortunate wound.

He had already turned the page when he realized he might have encountered something about the junkyard owner. He turned and shouted across the office: "Who's *Simp* Simpson?"

"Ichabod Simpson's son. *Only* son. Why?"

"Well I just read a funny piece about him and his car, and was wondering. Don't think I ever met him."

"I remember that. It *was* funny. You wouldn't've remembered him. He moved after his father died and the yard changed hands. Think he's in Missouri -- if he's still alive."

"Why didn't he stay at the yard, Herb? And when was it his father died, anyway? I mean, how much before I got here?"

"Just after Simp's accident, as I recall. Four, five years before you came, I think." The priest glanced at the edition in his hands. He had inadvertently picked up the sheaf containing events thirty-two years earlier, four years prior to his arrival, not two years where he'd left off. In the meantime, Herb had continued talking and the priest had to ask him to repeat his last few words.

"I said: Simp didn't want to work there and was happy to be bought out."

"Horace could *buy* it that soon?" The priest was incredulous.

"Never was clear. Talk at the time spoke of silent partners or a syndicate. The *bank* handled it. And Simp liked the price."

"Four years before I came, then. Thanks," he said and began paging furiously through the same volume for whatever he could find about Ichabod's death and its aftermath. He found several items two months later. *Senior Simpson Succumbs* was the huge headline.

Slow news week, or was he that important around here?

He continued reading. *Apparently important: Philanthropist ... much-missed ... friend to the community ... Methodist Church Overflows For Simpson Funeral ... survived by one son and his family; mourned by all Algoma ...*

59

He continued paging through subsequent issues, looking for news of the junkyard's sale. What he finally found was surprisingly brief, given the enormity of coverage for the death and funeral. The new owner was identified as *Horace Denver, a former employee of the junkyard.* A brief article three weeks later announced the loss to the community of Simp's family as they moved to west suburban Saint Louis. And that was it. Nothing more.

"Herb, what was Simp's real name?" he called to the editor who had disappeared around some boxes in the rear of the office.

"Don't remember. He was always called Simp," was the muffled reply. "Some said it wasn't just a contraction of his name. Now, I'm not saying he was dumb as dirt, but that accident should tell you somethin." He laughed and reappeared with an armful of galley sheets. "But, to be fair, he was a nice guy."

"Nice as his father?"

"No comparison! The old man was revered. Who else woulda taken in a stranger and let him work next to his own son? Course, *both* Ick and Simp treated the young fella with compassion. But what a surprise, I must say, when Horace ended up the new owner. If Simp had complained any, I think more eyebrows would've been raised; but he got his price and left, and the novelty faded. *And* -- Horace *has* done a nice job. Nobody'd deny that."

"So Horace was a stranger?" the priest asked coyly.

"Oh, yeah. Came from out west somewhere. Shows up of a sudden, and fore you knew it, he's at Simpson's."

"So, how *did* he pay for the yard?"

"Good question. Don't know who'd know for sure any more -- besides Horace." After a pause, he added: "'Cep, maybe Bob Lanner."

"The Bank President?"

"Yeah. His dad handled the transaction. Course, *he's* dead now. But Bob might know."

"Mr. Lanner's past retirement age himself, isn't he? You mean he wasn't running things at the bank thirty plus years ago?"

"Nah. His dad did till he died. Bob didn't get holda things till nearly his sixties. Hell, his dad died in *his* eighties still pretty sharp till almost the end. And stubborn! Proud too. Bob's a lot like him. The old man wouldn't think of givin up till he just couldn't go any more. Bob wouldn't, couldn't, ease him out. But he was a different man once he took over: in many ways nicer -- certainly surer of hisself. At times, though, he *can* be hard as his pappy. Can't say's I know who'll take over after Bob. But whoever it is, he'll probably have to shovel dirt on Bob, just like Bob did on his daddy. Probably Hugh."

Hugh Oster, current vice-president, the priest realized; not exactly a spring chicken either. "Mr. Lanner'd be seventyish, right?"

"No, ever bit of eighty; maybe a tetch more."

Talk about job security! He'd have to start researching Mr. Robert Lanner, President of the First National Bank of Algoma, Illinois. *But that'll be tricky. He's a primary source, as they say. Have to think a while on that.* But his work at the paper that day was clearly finished. So he tidied up and made for home.

CHAPTER IX

Algoma's courthouse, like many in neighboring counties, occupied an entire tree-ringed block in the central business district, and was one block off the main thoroughfare that was also a state highway. A small, sleepy county, its legal bureaucracy nonetheless churned out enough self-important paper work to keep a fair-sized staff on payroll. That bureaucracy was humming with busyness if not actual productivity as Father Wintermann climbed the steep, stone steps the next morning heading for the County Clerk's office and mopping his brow as he went.

Overnight he had devised an end run he was about to execute. The Clerk was a parishioner. Robert Lanner wasn't. If she could quietly find the legal record he wanted, he could avoid a potentially embarrassing discussion with Mr. Lanner. He found Dixie Duggan poring over estimates for electronic voting machines the Board of Supervisors had requested. She was rapidly concluding they wouldn't like the figures but, as a favor, the County Clerk was double-checking the charts for that evening's meeting, when the large form of her pastor suddenly filled her office doorway.

Dixie was a pert early-fortyish go-getter, and possessed of all the feistiness one might expect from a redhead. She also had a marvelous sense of humor and a magical smile, which she flashed as she looked up at the figure in her doorway. "What brings you here, Father?" she bubbled.

"I'm hoping you can help this legal layman find his way around your big bad courthouse," he smiled. "I want a record and don't know where to begin. By the way, that's a cute outfit."

"This old thing? Well, *thank* you," she said, smiling. "Marriage record? Be my guest. Got a million of 'em," she giggled. "Who you trying to track down?"

"Oh goodness no, Dixie. A property transaction. And from a while back too." He tried to look helpless.

"Got the wrong gal, Father. You want the Recorder of Deeds, Pat Putney."

"Well, I don't really know much about this stuff, and I was hoping an old hand -- *a veteran,"* he corrected himself with a self-deprecating smile, "-- could sort of walk me through it. Or at least guide my unsteady hand. You know ... "

"Sure. Let me finish this and we'll go over to Pat's and get you started."

Minutes later they were in the Recorder's office, and thanks to the staff, the priest had the information in no time -- with a minimum of fuss *and* without tipping his hand, especially to any pup-licker town fathers. His helpless waif image intact, he had bidden Dixie a grateful good-bye and was alone with his Xerox of the record. But he wasn't sure he was deciphering it properly.

Mr. Lanner was clearly the deal's agent. He also appeared to be the purchaser of record. But not the new owner! That was Horace Denver. A further surprise: the Lanner was not *William,* but *Robe*rt. Not the old man. The son! *This is getting curiouser and curiouser. I may just have to see old Bob anyway!*

All the way to Saint Helena's, and for another hour once he was again inside its air-conditioned quiet, he turned over and over in his mind what Herb Kuller had shared with him: that Lanner Sr. was a bank business bear. *But this is one dubious deal!* And *the old man's name is conspicuously absent. I can't imagine the son in this without his father's knowledge. Or, maybe it* was *outside daddy's awareness! Even so, why do it this way? And how'd he get away with it? To say it nicely, it's not straightforward. Why not be on the up and up? If Horace could afford to buy the yard, just write it up that way? And if he couldn't, why not a loan? But this? This was a bastard of a deal -- and for no apparent reason.* He had stumbled on precisely the right word, in fact, little realizing it at the moment.

He picked up his Breviary to clear his mind. But with Morning Prayer behind him, he was still making no sense of things. Then, out of nowhere, Maisie's name came back to him.

Most probably the dust on the end table got him thinking about when his cleaning lady was due -- *tomorrow,* he realized. *And just in time!* But *cleaning* then brought Maisie to mind *and* her unreturned phone call. Miffed, more at not remembering than at Maisie's dilatory behavior, he called her parents' number in search of their daughter again. Not unexpectedly, Maisie was cleaning a home. This time he asked where -- and set out to find her.

He was happy to learn that the home where Maisie could be found was not in sight of Annie's, though it was in her neighborhood. Small chance Annie'd be on her porch in this heat, but he wanted to avoid even a remote possibility of having to explain anything involving Maisie to the duchess.

The Chandler place was almost as imposing as Annie's, and belonged to a long established Methodist family. Father John had met them during his tenure on the Library Board a decade or so back, and it proved easy to free Maisie for a chat on their summer porch. By the time the two were finally alone he'd figured out just how to deal with the delicacy of discussing an employer with her employee, something further complicated by the town's unspoken racial code, and Annie's religious -- or irreligious -- adherence thereto.

Maisie Brown was a gentle young black woman from south of the tracks who had finished high school in undistinguished fashion and without further educational aspirations. She was employed, quite successfully, at several large homes and a few downtown businesses, including the bank. Her face and disposition were pleasant, and she was regarded by her employers as competent, sincere and self-effacing -- which Father John believed unfortunately meant she did her job and knew her place. He was also to discover that she truly was a sweet young woman, and possessed more acuity than townsfolk wanted to acknowledge.

"Maisie," the priest began, trying to put her at ease, "You may remember me. I've visited Miss Annie Verden. I'm her pastor, Father John Wintermann."

"Please to meet you, I'm sure, rev'ren." She began cautiously, holding a broom and dustpan tightly to her chest.

"As you probably know, Miss Annie's nephews visited recently. They took me into their confidence, indicating that their aunt has a breathing problem. They called it a bronchial something or other, as I recall."

"Yes, sir, that be right. I heard 'bout her bronichal problem. They told me I should be special careful fer her 'bout that. Even offer to pay me, you know, to hep be sure she take care of it proper."

"That's what I want to talk about. Miss Annie's health is important to me; I want to be sure she's doing what she should."

"Well, sir, those boys sure want that too. They done bought her all that ox'gen stuff she s'pose to use."

"So, Annie *does* have the oxygen," the priest played along. "A tank, I suppose. Large or small? What I mean is: will it last a while? I don't want her running out of something so necessary."

"Oh it be lastin a long while, I s'pose, the way she use it, and there bein a couple three tanks all told."

"The way she uses it?"

"Yes, sir. Not much at a time, I think. I tell her to, like the boys axe me, but bein there only onced every couple weeks, I cain't be sure she use it proper like in betweens. She want me to believe she do, fer sure."

"Well, Maisie, let me put it this way: I hope we can both conspire a bit. *I'll* think about ways I can get her to use the oxygen. Why don't *you* keep me posted about how she's doing? I mean, how her breathing is *and* how she's using the oxygen as well. Can you do that: help me to help Miss Annie?"

"I can be doin that, rev'ren. I can spire with you," she said, shifting to lean against the broom. She seemed to be breathing more easily now.

"Would you mind calling when you learn anything -- I suppose that'll be every couple of weeks, after you've cleaned there? I can give you my number."

"Oh, rev'ren, me and phones don' do so good."

That explains that! "Well, then, how can we do this? Let's see ... why don't I see you every so often -- perhaps where you work? Is there some place we could do that? What about here?"

"Well, I work the bank twiced a week. It be better there."

"Well, that settles it. At the bank, then, which *is* a better place. But before I go, you mentioned more than one oxygen tank. Just how many does she have? And are they the same size?"

"'Zactly three. Like I said, rev'ren. 'Bout this big I s'pose," and she stooped to gesture with her hand two feet off the floor. "They all be that size."

"Small tanks then. Ok. So, why not let me know when each one's empty. That'll tell us if she's using them and how fast; and when she gets low -- you know, down to the last one -- I could do something to help her order more."

"That be easy. I be glad to hep."

"Maisie, you're precious. I appreciate your willingness to assist a lady like Miss Annie," the priest said as he stepped off the porch. Then he turned to add: "Could you be available to clean my rectory?"

She shifted from foot to foot and wrinkled her forehead. "I don't mean to say I be busy." She paused in some internal confusion:: "I be glad to consider it, but I gotta hafta fit it in. I might could do it if you's to be flex'ble."

67

"Oh, I'm sure I can be. and I'll get back to you, Maisie. Thank you; very much." And he went down the stairs and around the house to his Taurus waiting in the shade of the big elm that overshadowed half the hot street.

He needed to take mental inventory. So he drove to the drugstore, parked, and stepped from the sizzling outdoors into its shaded, ceiling-fanned interior. He sat down in a booth, pulled out a small pad and a pen, looked up at the woman behind the counter and smiled pleasantly. "Hi, Frieda. Got lots to think about. It'll take a whole banana split with gobs of whipped cream to get through it. Take your time; no hurry." And he smiled again.

So did she. "I've got just the thing to handle it. Want some water too?"

"Yes, thanks, dear" he said to one of his favorite pup-lickers as he bent to the task at hand.

Frieda was half the ownership team, a large woman who never met a stranger. Her pharmacist husband would soon appear at the cage window once he heard a familiar customer's voice, Father John knew. He also knew that one or both of them would join him for a little gossip. He wrote fast to finish his jottings confidentially.

Frieda meanwhile was happily crafting a gargantuan creation for a favorite customer. She was a big-boned, jolly woman of Scandinavian descent, and a staunch Lutheran at Pastor Thurman's church. She and Fred knew everyone in town, Father John was convinced, and many things about them as well. Their drugstore was one of his regular stops. The ice cream was good, the company genial, and the gossip free.

Let's see. The nephews say Annie has health problems, which I can't corroborate. Is Annie cooperating with them -- even after they spent money on oxygen plus a fan, to boot? I should ask if Maisie actually sees Annie use that oxygen. And those questions about gas in Carbondale! If they were going to get oxygen, why ask about that, especially if you're a chemist? And Horace: he owns the yard -- outright or what? That title transfer has a strange odor!

He'd been scribbling furiously and looked up to rest. Just then another customer came in and went to the over-the-counter medicines. He began writing again. *Annie helped Horace get a job; now he owns the place. Annie's very protective of Horace, but if he really owns the yard he shouldn't* need *that. Maybe she's trying to fill a void in a lonely man's life -- or her own ... No, not the latter. Also, still don't know exactly how or why Annie and Horace got connected.*

He looked up and saw Frieda coming, pretending to stagger under the weight of the ice cream. They both laughed as she eased it with mock difficulty onto his table, and then, sure enough, squeezed in opposite him. He put his notes safely into his shirt pocket and reached for the water Frieda had also brought. There were two more glasses, for herself and, he was sure, her husband, who was just coming after accepting a lady's money for some remedy or other. He pulled up a wire-backed chair and sat at the end of the booth. Both were eager for the latest the priest might have.

"Slow news day?" he asked, and smiled.

"Sure has been," Frieda acknowledged. "Nothin much shakin."

"Business is slow, too," Fred added. "What about your day? Quiet also?"

"Not really. But not all *that* exciting! Got some roasting ears from one of my farmers last weekend. Field corn, you know. But still pretty good. Sweet's just coming in, I'm told. Can't wait; love that stuff. Anybody new sick in town?" the priest said before taking his first gulp of the gooey concoction. A little fishing expedition! Maybe Annie'd tipped her hand to them. Or the nephews may have gotten the oxygen from them.

"Nothin earth shattering, for sure," the man said, looking at his wife, who confirmed the verdict with arched eyebrows as she downed some water. "Why?"

"Nothing special. Just curious! I went over to the Smile yesterday to browse through some copies from before I came to town," he said, with his mouth half full of the wonderful dessert. "Little bored lately, I guess. Anyway, there's stuff like that in there, you know. I did hear the Methodists are thinking of changing their October Mulligan to a Trough." That should buy breathing room.

"Why would they wanna do that?" Fred asked with some emotion. Town pharmacist for years, he prided himself on monitoring the town's health, emotional and social, as well as physical. This had eluded his antennae, and was important, at that. "They got a darned good stew recipe. Besides, a big outdoor fire in late October can be dicey. Not to mention the cost of steak! They'll have to charge more!"

"Don't see how to avoid it," the priest agreed. "Guess they want something different. Guess, too, they figure traffic'll bear a price hike." After a reflective pause, he added: "But it'll take some convincing for my parishioners to be as supportive as they've been. Ecumenism's one thing -- a price boost is something else, I'm afraid!"

Fred still couldn't believe it. "Would you work a fire like that, Frieda?" he asked. She nodded a firm negative. "Me neither." They all knew a late October in-ground fire bed was tempting fate. The least bit of rain, and keeping the coals white-hot for the steaks they'd lay directly on them would be next to impossible. "Gonna put it under some sort of lean-to?"

Frieda chimed in: "Have to. Which is another expense. Think they'll use beef or pork? Or real buffalo meat?" She was alluding to the origins of the practice. Plainsmen had used the method to cook for hunting parties, often as not, bison they'd shot themselves. And while Father John knew all that, he'd never been able to find out why they had called it a Trough, or why something that rhymed with bow was spelled that way. Maybe it was French in origin.

"Surely beef -- not?" Fred was saying. "Where they gonna get enough buffalo steak?" He had sounded their consensus. "Gonna ask me the first Methodist I see what's going on. This is somethin, all right. Just can't figure it."

Father John was still devouring his ice cream and Frieda was about to share a new tidbit, when two youngsters appeared in the doorway, clamoring for *two Skis to go*. "Hang on; I got somethin else," she told the two, and slipped past her husband to draw the two drinks. As she headed to the soda fountain she looked over her shoulder and winked: "Gotta get 'em their sodie fix."

Ski was the soft drink of choice for youth throughout the county. It had more caffeine, and was the odds on favorite of the younger set. The Becker Pharmacy had added it to their fountain array to court the youthful loyalty jeopardized when the malt shop reverted to pharmacy status. So intense was product identification among area youngsters that the company did repeated market analysis to figure out the anomalous regional popularity. That business decision had proven shrewd beyond ordinary prognostication.

As soon as the front door slammed behind the boys, Frieda shared her news, well before she resumed her seat. "Have you heard about the art show?" This was apparently aimed at the priest, as her husband didn't react.

He hadn't. And with his mouth full of ice cream he could only flash an encouragingly expectant look. She continued. "Well, the Library Board decided to hold a local-talent art show to raise money. Before school ended they got cooperation from both elementary schools, and the high school too. Now they're asking adults to submit all sorts of things -- crafts as well as artsy stuff. Sure you haven't heard?" She was looking directly at Father John, expectantly.

"Well, now you spell it out, I remember hearing about the school part," the priest said, between bites. "But I didn't know they planned to ask us oldsters too. I'll have to break out my crayons!"

"Well, now," Fred chimed in. "Frieda and me have in fact been thinkin you'd be the perfect person ..."

"No, you don't," the priest cut in quickly. "I haven't one ounce of that kind of talent. And don't intend to develop any."

"No," Fred continued. "Not to submit anythin. *To help organize and promote!* You'd be wonderful. You used to be on the Board. Why not help put this over? Everyone knows you, and you can help get the reluctant artists."

"That's different. But I'm still not sure. I did my time, and the current people are good folks." He hoped that would end the discussion. It didn't.

"No, Father. We've talked this over quite a bit. I mentioned it to one or two Board members -- didn't think you'd mind. They say it's top-flight and wanna sign you on." Fred was nodding at his wife's sentiment.

Not anxious for an argument, Father John opted for a gracious stall tactic. "I *will* think about that. *Very* carefully." And he smiled, trying to end things.

"Goodness. Must get back," he said, to decidedly forestall anything further. "What's the damage here?" he asked Frieda, though he knew well his favorite delicacy's price. Handing her three dollars before she could answer, he waved for her to keep it all. And rising ponderously, he made for the door.

"*Do* think about it, Father," Frieda said, as her husband nodded encouragingly behind her.

"I will. But don't hold your breath just yet." And he gently let the door close behind him as he stepped out of the quiet, shaded cool of ceiling fans and floor length blinds into the oppressive blast furnace.

Do I need this? I'm not inclined to think so! Still, maybe I could offer a kick-off prayer. It is interesting. I wonder how much talent there is in town. And will they get many crafts? Yeah. Probably crafts! He started the engine to return to the isolation of his study, where he'd work on next Sunday's homily.

CHAPTER X

Next morning, something about the way the light struck the lobby wall as he was shutting the rectory door after Mass caused him to pause and stare at Horace's painting. He had to reopen the door onto already bright sunlight to recreate the effect, and once the cruel glare fell on the wall again, he was certain he'd seen something peculiar. The picture seemed to bulge slightly, and when he ran his fingers over it tentatively he was convinced of it. He took the picture from its hook and brought it into his office to examine more closely. He realized with some chagrin that it was an oil not a print, as he had surmised all these weeks, rather delicately painted at that. Furthermore, it was signed illegibly in a tiny scrawl at the right bottom. *Probably not a famous artist! But what could that bulge be?*

The reverse -- plain, stout poster board -- offered no answer. He delicately pried the picture from its frame, removing the many brads, and gently lifted it out. Between the backing and the painting was an envelope addressed in a delicate hand to Horace Denver and containing five crisp ten-dollar bills plus a brief note in the same small script:

Horace – the painting may have some value – treasure it – use the money for yourself

No signature, no final punctuation, not *any* clue as to the sender. Horace had mistakenly given away his own gift, probably not identifying it as a gift in the first place. *But, from whom? Maybe the artist is better known than I thought!* And the money, besides crisp, also had sequential serial numbers. *Obviously purposefully acquired from some financial institution.*

There was no debate. *I'll return the picture and tell Horace the unvarnished truth: I simply can't keep it in good faith. I hope Horace isn't offended, since his gesture of several weeks earlier was a genuinely grateful one.*

Driving across town he thought the note's handwriting faintly familiar, but he was more interested in restoring to Horace what was rightfully his than in puzzling that out. As he drove up, he was relieved to see Horace inside the fence. There'd be no confrontation with Dog.

Their conversation was interesting, if predictably terse. Horace appeared more relieved than flustered as the priest explained that the donor would certainly want Horace to have it. Nor did the junkman offer anything in its place. After Father John told him about the envelope and that it was back in its original location, the junkman merely mumbled *thank you* and added something about the giver being relieved. The priest wasn't certain, but he thought he heard him say *she.*

As he left, John Wintermann realized he was no closer to cleaning up this tangle of mental underbrush. If anything, it was now messier. On impulse he decided to take the day off. Perhaps that would clear his mind and help him see things from a fresher angle.

He'd call Harold. They could drive south, visit another priest or two, see some of the beauty of the Shawnee National Forest, and share a meal or two besides.

Harold didn't answer. *Plan B. Go by myself. Less desirable, but under the circumstances ...* The more he thought about it, in fact, not very desirable at all. He called Jim Witten in Herrin, who would gladly join him for lunch but was unable to get freed later. It was a start; he'd make the rest of it happen somehow. So he set off south, avoiding Interstates to look at the scenery along the way.

Driving gave him time -- he often did that -- to think. His first priority was to revisit the day's possibilities, which he did between glimpses of the lush but thirsty cornfields and occasional stands of timber. The whole way his windshield was glazing over with the corpses of flying insects. The region is home to hundreds of species, probably thousands, all thriving on the abundant vegetation in this fertile southern piece of the Corn Belt. *Maybe a film in Carbondale; or else the Garden of the Gods outside Harrisburg! And who else might be available? Maybe Ed Wallen at the Newman Center. I'll call from Herrin. Good enough for now; I can always adjust if other ideas come. It's a day off! Don't be rigid!*

He began to think about Jim and his great uncle, Right Reverend Monsignor John Witten, one of the legendary German priests of the diocese. Stories about the mainly German characters dominating the presbyterate in the early decades of the century suggested they were often lovable autocrats with outrageous senses of humor.

Witten's uncle had been a huge man, while the bishop of his latter days was small of stature. He loved to literally talk down to his religious superior. Hovering over the man, he'd speak in exaggeratedly respectful tones, a posture that had to be -- was no doubt meant to be -- physically intimidating. Story after similar story evidenced an inside joke among those German movers and shakers. They hadn't expected the smaller man, German like themselves, to be named bishop, and they found many ways of saying so, of keeping him in his place within their ethnic ranks. A further irony was that the tiny bishop struck awe, if not fear, into the hearts of everyone else, clergy and laity alike. But not his Germanic confreres.

Jim was nothing like his great uncle. Quiet, unassuming and much loved by parishioners and fellow priests alike, he was, if anything, too easily put upon. John suspected his unavailability the rest of that day to stem from yet again being unable to say no to something. But they'd enjoy lunch together, he was sure, at one of the town's Italian restaurants. At least he hoped so.

As he neared Herrin, he remembered it's violent Depression era history. The twenties and thirties spawned much national sadness, and Southern Illinois had endured its share: violence at area mines; the ascendancy of the Ku Klux Klan with their demonstrated hatred of Blacks and Catholics; and the ambivalent presence of Charlie Birger.

Folklore painted the Birger Gang as Robin Hood types, but they were simply gangsters, though colorful and anomalous in their own way. They hated the KKK and once openly protected a priest at a high school basketball game in West Frankfort. Nor was that their only Church connection. They insisted that injured gang members be treated at Belleville's Catholic hospital. And they once called Herrin's Catholic school to have it dismissed early. Wisely enough, it was. Though they hadn't said so, the gang didn't want the children harmed when they blew up a garage across the street around 2:30 that afternoon. The last of them was hung in Benton in the mid-thirties, a replica of the gallows, in fact, remaining today as a town monument. The gang was one of the reasons the county earned the dubiously distinctive nickname *Bloody Williamson.*

Just outside Pinckneyville he thought again of Horace and the painting. *Might he have a lady love? Horace, a girl friend? Not likely! Anyway, no relationship in Southern Illinois would ever allow the lady to give money to her man. Perhaps it's Annie, or another of the town's senior set taking pity on him.* But try as he might, he couldn't think of anyone in Algoma with an eye for paintings like that. *Maybe I should hang out at that art show to see if any suspects emerge. When is it again? Two weeks? Yeah. I can do that.*

The Fourth had come and gone, the heat once again keeping him away from the fireworks. As in other years, he watched briefly from his bedroom window. Each year it grew longer, noisier, glitzier. Children loved it, hardly minding the temperatures and mosquitoes that kept the likes of him away. The library event was two weeks away yet, between the Lion's Club Homecoming and the early August Lutheran Bazaar. He could endure two more weeks to find out what might emerge from that high society moment. He'd have to.

The day off was therapeutic. The Italian lunch was enjoyable, and after some shopping at the Carbondale mall, the late afternoon movie with Ed was the right comic relief to wash away any remaining stress. Diocesan scuttlebutt not supplied earlier by Jim, Ed seemed to have, and he also described the latest college fads over a light but exotic evening meal at one of the town's ethnic restaurants, ending what Father John decided had been a most relaxing day off. And classical music on his radio all the way home allowed him to avoid mentally rehashing parochial concerns. Time enough for that tomorrow.

CHAPTER XI

The next two weeks were flooded with parochial tasks. Only occasionally did Father John's mind turn to Horace or Annie. But there was nothing new on those fronts, nor did he even see either of them, the weather being much too hot to catch Horace in his usual haunts or for Annie to venture out to church.

Annie had her name in the paper once, in connection with the art show. She and members of several prominent town families were listed as donors of *family heirlooms*. Father John wondered who'd buy all these treasures. He supposed it would be like most parochial charity events: the same people who donated materials, not to mention time, would spend money to buy everything back. And feel like a good -- and charitable -- time was had by all.

When the day came, however, the priest was surprised to see numerous strangers in attendance, some with cars sporting even Missouri and Indiana license plates, and at least one from Kentucky. The committee had evidently done a good publicity job. He'd managed to keep off the committee without hurting feelings, and it was a relief to have no official duties. But he was one of the first to arrive the day of the sale and, as it happened, one of the last to leave.

Many items were priced outright, but a surprising number had been set aside for auction, which part of the program wouldn't begin till two o'clock. He made a late morning purchase: some hand-made doilies for his nieces. It served to demonstrate loyal support as well as mask his extended and over-interested presence. He originally intended to break away by noon until he noticed not one, but two oils in the style of Horace's, signed in the same small, illegible scrawl. Both were to be auctioned. And the donor wasn't listed.

Late in the afternoon two different outsiders, ladies both, purchased the oils. He engaged the more approachable one in conversation shortly after she paid. "Who is the artist? Of the piece you bought? If you don't mind my asking."

Eyeing his clerical collar, the woman answered a bit aloofly, if politely: "I don't believe we've met."

"Oh, forgive me. How thoughtless. I'm Father John Wintermann, the Catholic pastor here. It's just that I've seen only one other piece by that artist before today, and haven't been able to find out who it is. I may be interested in one, if I could discover where to see more of them."

"It's nice to meet you, Father," she said, warming slightly. "I'm Doreen Smithfield from Saint Louis. I must say, I'm not sure about the artist either; I can't decipher the signature. But I'll ask. Someone here must surely know."

"How did you hear of our event? Saint Louis is far enough away. I'm surprised we could draw any of you here," the priest said amiably.

"Well, I follow these things, and your group was clever enough to advertise in a specialized journal. Though, in fact, I heard about it from a friend who had read the ad. She purchased the artist's other painting, by the way. We came in the same car, you see. Would you like to meet her, see her oil? I saw you looking at this one earlier. I thought you might be bidding against me."

"No, I've seen both paintings, thanks. I really can't afford what you ended up paying. Good for the library, but bad for my pocketbook, I'm afraid. But if I could learn the artist's name, and if he or she is alive, approachable ... you know ... perhaps I could negotiate a price I *can* afford."

"Well, let me ask. I'll find you before we go and let you know."

He thanked her and stepped away to make small talk with one of the committee members beaming nearby, dollar signs floating just behind her light blue eyeballs and nearly visible each time she blinked. She figured the event had done much better than anticipated, and he agreed.

In a few minutes the Saint Louis lady returned with a quizzical look on her face. "They weren't sure about the artist either. But both oils were donated by a local woman, who didn't furnish that -- nor did the committee think to ask for it, it seems. They promised to find out, however, because we'd both like to know," she said, indicating her traveling companion. "You could, no doubt, do the same on your own. Do you know a Miss Annie Verden?"

He sufficiently concealed his amused surprise, he believed, and replied: "Yes. And I can certainly satisfy my own curiosity by asking. So, there's no need to contact me. But are you quite sure the committee will get that to you? You wouldn't want me to call you once I find out, would you?"

"No need of that. However, on second thought, perhaps -- just to be sure -- I *could* give you my phone number."

"We can call each other as needed," the priest said. They exchanged phone numbers and he jotted the lady's hastily in his shirt pocket notebook. She thanked him for his thoughtfulness, and he told her it was nothing. As the two ladies left he felt some relief. He now had a more plausible excuse for approaching Annie. And he now also knew where Horace's painting had come from. *Did she expect him to hang it in that shed of his? And why the cash?*

It took another day to get into the proper frame of mind for Annie. He tried to remember if he'd ever seen any such oils in her home. The two were of a distinctly dreamy style, mostly bright pastels. Even Horace's Last Supper had looked like that. They were distinctive enough he'd have remembered, he believed, so he was fairly certain he hadn't seen anything like that at the mansion. Perhaps they'd been upstairs or even in the attic. He also wasn't sure any more of the authenticity of the report as to their being *heirlooms.*

He decided to visit Bob Lanner first. The past two weeks had been too full to tack that loose end down, and he felt it easier to tackle it first. Late morning seemed optimum. The lunch hour would provide a perfect excuse, should the moment prove difficult, as he fully expected it might. He knew he'd uncovered a sticky situation and wanted to be ready for a very defensive bank president, though he was puzzled why no one else had ever raised similar questions. Perhaps his suspicions were unfounded. Anyway, he'd know soon.

He had met the banker numerous times socially, but didn't know him well. What he did know was that the man's age belied his mental acuity -- even now the priest continued to hear of his shrewd financial ability and the occasional ice-cold business maneuver. He also came forewarned about the man's charm. In his eighties perhaps, but not to be taken lightly!

When the priest walked into the banker's office late the next morning, the man didn't rise from behind his large desk; but his voice was full of sweet good humor. He greeted the priest overly politely and offered him a seat in one of the plush chairs directly in front of his desk. "Do rest a bit from the heat, Father. Are you here for financial assistance or just payin a social call on an old man?" He was dressed in a gray suit and spoke with a faint southern drawl. His smile seemed forced, but only slightly.

"Mostly social, I think. How are you? Staying where it's cool, I hope."

"Fine, 'cep for my sugar diabetes. Actin up a mite! The price you pay for gettin on, I'm 'fraid. Just have to limit my pie *and* late afternoon nippin. God, how I enjoy a bourbon and branch! Ever so often I just have to cut back; and this week seems to be one of them times. But, nothin big to worry 'bout, really."

"Sorry to hear about the inconvenience. Myself, I was at the library gala yesterday. I noticed the bank was a sponsor, and I want to commend you. Marvelous idea! Why didn't someone think of it before? And it was attended far better than ever I dreamed. Shows you what I know about that. I'm sure they'll be happy with the results, and I for one can't wait to hear the report in next week's Smile. So first off, thank you for your assistance with that."

"Well, that's kind of you. We're happy to help with those sorts of things," he said, the words lazily slipping through his lips

85

with a slight lilt and distinctly southern sound. Shifting in his chair, he cleared his throat and continued in his more familiar, *business,* voice. "Truthfully, it wasn't all that much, and certainly not very difficult. Be willin to help with more things when people dream them up … *and* if they think to ask. Be surprised how many times we have to approach people to see if they want help." He seemed to puff up with the ingenuity and pride of it all, and reached for an unlit cigar resting in a large ashtray beside his desk pad. The priest couldn't help noticing how bare the desk was. Except for the ashtray, there was only a small American flag on the right front of the desktop and a phone at the banker's left. "Wasn't able to get to it myself," he said as he began chewing on the still unlit cigar. "Well attended, you say?"

"Very much so. Brought in people from Saint Louis and farther. Besides helping put Algoma on the map, it's nice to have outside money for things like the library." He smiled, sure he'd sounded a theme to warm any banker's heart. When he heard *that's very true,* he congratulated himself on trusting his intuition. Buoyed, he launched into his real agenda, but on tiptoe nonetheless.

"If I might change the subject. You may've heard some weeks ago that Horace Denver had a nasty accident. Well, I was called to the hospital, and to make a long story short, Horace stayed with me a couple of days. I must admit I never thought much about him before, but after he stayed with me and I got to know him a bit better, I began wondering about something. If you'll pardon my curiosity and impudence, I frankly began to wonder how he acquired the yard in the first place. He certainly isn't a man of many words, but he did tell me he was an orphan who came to work for old Mr. Simpson when he first got here."

"Now, all that was before I came here. Even though I've been here a while, Horace got the yard an even longer while ago -- and I couldn't get over the fact that he was able to buy that yard after working only a few years before the old man's death. I was wondering how an orphan could do that. Is he some sort of financial wizard? Doesn't strike me as the type." Throughout, the priest had been watching for any signs the old man's face might betray. There were none.

"Well, that surely *was* a long time ago. My daddy was Horace's mentor, so to speak. I'm not sure I recollect all that much 'bout it." The banker's face still had not changed an iota. The placid, even friendly, countenance hadn't flickered, and the small smile still lingered on the edges of his lips.

"Oh, really?" The priest paused for effect. "As I understand it, it was yourself who helped him with the matter." He debated just how to phrase the next sentence. Fortunately he chose the right words. "These are serious, even, shall I say, delicate, matters, Mr. Lanner. I trust I did not misunderstand."

Lanner's face now changed dramatically. He put his cigar down and frowned a moment in silence, swiveling his chair away from the desk so that he no longer faced the priest but was directly gazing out the window to his left. The priest allowed him the next move, which didn't come for long minute or more. When he swiveled back, he asked: "How much did Horace say?"

The priest paused only momentarily and decided to play his cards close to the vest. "More than I wanted to know." *That ought to prime the pump!*

The banker slowly reached for his phone, pressed a button and spoke quietly into the receiver: "Bobbi Sue, hold appointments and calls; I'll be a good while longer heah." Then he turned back to the priest and sighed deeply before continuing: "You're right to call these matters ... *delicate.* I know priests are used to delicate and private matters. As a banker I am too. I must be certain this conversation stays within this room. Is that your understandin as well?"

He had struck gold. *One vein or the mother lode? Should know shortly.* "You can be certain it is, Mr. Lanner. Why else do you think I've spoken so carefully? I mean: I owe that much to Horace, do I not?" Nothing untrue there, but he astonished himself at his ability to improvise.

"Well, it's a long story -- you must know as much already -- and one I've told no one outside my family. Would certainly not be discussin it with you today but that you've come upon it. It's ironic that it wouldn't come to light till nearly sixty years later." Mental arithmetic told the priest it was the mother lode: Horace's local history wasn't that old. *What in the world have I stumbled onto?*

"Horace didn't know about his parents beforehand, but my daddy -- and I -- believed he came to know a short time after arrivin heah." The banker flashed a look at the priest as if to ask whether he agreed. Father John gave a small smirk and tilted his head abruptly a degree or so to the right in a noncommittal gesture he hoped would encourage the man to continue. Apparently it did.

"At the time Ich Simpson died, my daddy was aware of several things. He knew his son, Simp, didn't want to work the business. He also knew Ich liked Horace and wanted him to continue workin at the yard -- *maybe even,* in lieu of Ick's own boy, to have it someday. And my daddy knew that Ichabod didn't figure on havin to settle those matters for some years -- he thought he'd have time to work out somethin to take care of both his son *and* Horace."

"But one thing daddy didn't know for sure, yet was strongly convinced of: he came to believe that Horace was shrewder'n most people thought, and that he'd figured out his parents came from Algoma. In fact, daddy believed Horace knew or was close to knowin *exactly* who they were. That worried my daddy."

It was now the priest's turn to control his own facial reactions. He didn't move a muscle, just kept sitting dead still with a serious look on his face, hoping to encourage the banker to go on. But his mind was racing in an effort to assimilate what he was hearing, without missing anything new as it poured in.

"My daddy didn't know exactly how, but he became convinced he was gonna have to buy Horace's silence. That meant gettin the junkyard into Horace's hands -- and without Simpson's knowledge -- Simp, I mean, because this'd all have to happen if and after Ichabod died. God help us if the old man simply wanted to retire! What my daddy was puttin together couldn't be pulled off unless Simp was the only one involved. Old man Simpson wouldn't have gone for it. But Simp would be easy. As, in fact, it turned out he was."

Still the priest sat impassively. And the banker continued. "Well, Icahabod's death was sudden, all right; and -- pardon my soundin insensitive by sayin it just like this -- in a way, *convenient.* But it was sooner'n my daddy would've liked. Nonetheless, he put his plan into motion when he had to."

"A day or two after the funeral he called Simp in for a fatherly, compassionate talk. He was all sympathy and concern, and spoke of several confidential discussions he'd had with Ichabod -- all non-existent, mind you -- about the junkyard's future. He said Simp's daddy realized his son didn't want the large responsibility of the business -- hell, ever'one knew that -- and yet he wanted his son to be compensated properly for all the hard work that he, Ichabod, put into the yard -- which lots of townsfolk had also come to realize. Weren't nothin in that -- 'ceptin for them *private conversations, which wasn't real at all* -- that most ever'body in town didn't already know."

"Then my daddy sprung the rest of it on Simp, who never suspected a solitary thing. He told Simp his daddy had begun to put money aside for Horace with us at the bank -- wasn't true; but could'a been. Wasn't all that much yet, he said, but with interest and all it was just over thirty thousand dollars. Not near enough to buy the yard, a course, but a start. And he said we were encouraged by that; and encouraged too by how hard Horace worked; and feelin good-hearted as we did ... how we was gonna loan Horace the rest of the buyin price, and we just knowed he'd be good for it in proper time. *How does one hundred and forty thousand dollars sound? All in one lump sum. All for you, Simp.* That's what my daddy said. I know. I was there. I was in on that conversation. There was just the three of us in this heah room: Simp, daddy and me."

"Now, my daddy figured Simp might not jump at it all at once, and daddy knew he could go higher if he had to. Well, sure enough, Simp took a deep breath, said how grateful he was and all but that he'd have to think bout it, a course -- talk it over with his wife. Couple a days later he was back sayin how's he didn't wish to appear greedy or anythin, but the figure sounded a little low, bein's how he was gonna have to live on that. And my daddy was the soul of concern; he agreed it oughtta -- and could be -- higher. We settled on two hundred thousand. Simp walked out a happy man and left town a week or two later. And nobody ain't never heard nothin from him since."

"My daddy turned to me after Simp left this office that day. *Cheap at that,* he said, what with our family name intact and all. And I agreed, and told him so. And thanked him. He told me it wouldn't be all that big a bite because he had himself begun to put money aside from the start just in case Horace ever showed up to ... blackmail ... us. Daddy didn't like the word, but used it anyhow. Hell, once we got to really know Horace, we knew he'd never do *that.* But in his own proud way, daddy felt all along we owed him somethin like that anyway."

The banker continued speaking in his quiet, soft, undramatic drawl. "It was all just too perfectly right, you know. Horace'd never talk; he'd be doin a job he liked doin, and one the town needed; and we did owe him that -- or somethin, anyway. So Horace owned the junkyard outright -- from that moment on! Never was a loan, a course. Horace was told he'd inherited the yard. Far as I know, he don't think nothin different today. And, yes, I worked out the legalities with him. Me; not daddy! It was daddy's idea that I handle it start to finish. It was rightly my responsibility, so to speak."

The priest was beginning to feel numb. He had said that he'd been told more than he wanted to know. In fact, it wasn't Horace but the banker who was only now doing that. He continued to listen; and the banker continued talking. It was probably therapeutic.

"I thought the hardest thing would be facin him if he should ever show up in my life. Turned out, that wasn't hard at all. I mean, enough time had gone by; and anyway, I never knew him till he did appear heah. And Horace isn't the kind of man to feel -- you know -- sentimental -- *or* pushy, either one."

"But Horace's momma, now! That's another story! That was never easy. From the very start it was difficult. Somethin 'bout motherhood, I guess. She was troubled that she had to give him up. She'd cry over not knowin how he was growin up, what he was turnin out to be like. All those things I couldn't get excited about. But she could. And did. She knew where he was, a course; we both did. But like myself, she didn't think it wise to inquire directly. So she said. But I have a suspicion she stayed in touch with that orphanage anyway -- sent donations from time to time; that sort of thing. Could never be sure, but I suspect as much. And, a course, she was overjoyed when he showed up heah. For all I know she maybe even enabled that too."

"Oh, I tried to help her from the start; financially. That was a mistake, my daddy tried to tell me. He was probably right, because I couldn't shut it off. Can't, even now. God, how long's it been? Near sixty years or so! Off and on, payments; always small; always cash. They were easy to conceal from my wife, once I married; pocket change, really, most of the time."

"Thank goodness the lady was never greedy; never set a price; never said it had to be such and such an amount at *any* time. And she was always so grateful -- like it was my idea, 'stead'a hers. I guess that was right. *Was* my idea! Thing is, don't think she could've lived without it -- not even now. Although, I do get just a touch upset ever once in a while -- just a touch, mind you -- and only for a time. It don't last long, that feelin don't."

"But, I mean, in some ways, sixty years ought to be way long enough! She can't seem to curb her spendin. Lives like the money won't never end." He paused. "And I guess it won't. Can't. Though, lately there are some signs ... " His voice trailed off, and, for the first time, just momentarily, it seemed there might be nothing more he could think to say.

But there was.

"We just couldn't let the town know; ever. Two highly visible young people from the best of families! You understand, I'm sure," he said, glancing up at the priest. It was as though he had been talking to himself the past ten minutes.

He didn't wait for a response: "I must say, her family was superb; mine too. Never a word of recrimination to either of us! And when it came clear what had to be done, they arranged for her to *visit friends and relatives out West.* Had to be far away, you understand. Fact is, they constructed a 'laborate story about all the places she was to visit on this extended cultural trip. And to her credit, after she had the baby and placed it with the Sisters in Colorado, she did, in fact, travel to a few of those places just so's she could talk about 'em later."

"She and I never thought about continuin our relationship. Hell, it wasn't much a one anyway -- not all that long, I mean -- before the pregnancy; and she was gone near a year. But most of all, it was the religious differences. Would never have worked, both families felt. I think I agreed, even then; certainly did later. I'm not sure what she thought about that, but she surely never suggested we resume anythin. I think she just wanted to forget about that piece of it. But her boy, now, he was always on her mind. I imagine she's come to some peace now that he's been here all these years."

The priest realized that he still didn't know who the mother was. But neither did he want to -- or need to. What had begun as a fishing expedition about Horace and the junkyard had turned into a regular big game hunt, with trophies he didn't want to display -- or could, had he wanted to. He took advantage of the momentary silence to do some needed people-patching.

"Mr. Lanner, I can only imagine how difficult this must be to unburden yourself to a virtual stranger. I just hope that letting go of this will truly make your load lighter. And I want to assure you of several things. Horace didn't send me here. Nor will I ever share any of this with anyone, least of all with him. I hope you feel assured I'll honor the sacredness of what you've shared, and its private and special importance. I don't wish to make your life more complicated, more shame-filled or sad. I must say, in fact, that what you and your family have done for Horace is extremely kind and loving. And, if I might add, that's reassuring to an aging cleric who is more and more susceptible to cynicism as the years go by. *And finally,* I'm humbled, as I always am, when people take me into their confidence. Thank you for telling me everything so forthrightly. It *is* safe with me, I promise."

It wasn't surprise so much as relief that spread over the older man's face. He rose ponderously and spoke: "Thank you for listenin, and for listenin so ... " He was searching for the precise word: " ... so calmly; without judgin. Pardon me, but I didn't expect that from -- well, from anyone in Algoma, I suppose, but especially, if you'll forgive me, from a priest. I just didn't think ... "

Father John got up and walked around behind the desk, extended both hands and, clasping the banker's right hand within them, broke in on his sentence: "Having a child is a wonderful thing; doing well for that child is even better." After firmly pumping the old man's arm, the priest added a final reassurance: "*I really don't want -- or intend -- to share this with Horace.* That's for you to do, if ever it's to be done. Thank you again for your time, and for your straightforwardness. I must be getting home for lunch," he said, looking at his watch. The noon meal had turned out to be the perfect excuse. As the priest was leaving, he saw the old man's face begin to lose its look of pain. He hoped that signaled the advent of some personal peace for him.

There was vastly less to the Horace mystery now, just the part about his mother. All he knew with certainty was that she was still in Algoma and wasn't Methodist. That was enough. He already knew more than enough about his father. But still, some things about Horace -- and Annie -- made no sense, including the painting. And the fifty dollars. *More work ahead.*

CHAPTER XII

"I admired two lovely oils at the Library Affair, Annie, and was wondering who had painted them." Father Wintermann was at the mansion the very next afternoon, a man on a mission, and he found himself sitting with the duchess under the gentle breeze of her summer porch fan, waiting for tea to steep. "The committee didn't know, but said to ask you, since you donated them."

Annie's reaction was interesting. With no visible shift of emotion she said, "Why, Father John. I didn't know you had an eye for beauty." Blinking coquettishly, she made it sound almost risqué. When he didn't reply, she continued, "Did you really admire them that much?"

"Why, yes, I did."

"And what did you like most about them, especially?"

"Great vibrancy of color but also a certain dreaminess. Something like the Impressionists, but more modern. I can't explain it well, I'm afraid. But I liked them. The price they went for was far too dear for me. But I'd love to have one like them, I believe ... if the price were right ... something I could afford. You *do* realize that what priests are paid ... " he searched for a delicate phrase, " ... could hardly be called *dear!*" He smiled.

She smiled back and excused herself. She returned momentarily with a steaming kettle on her large tray, and all the other accouterments for high tea as well. She settled back onto her padded chair as the priest did the honors: lemon, sugar, cream: the works. Only after a satisfying sip from her cup, did she speak again. "You really liked them, then, the paintings?"

"Oh, yes -- as I said."

"Did they remind you of Horace's Last Supper?" She paused, and added: "He told me about his faux pas."

Faux pas? A small oversight! "Oh," he said, feigning faint recollection and hoping she'd believe that. "Was that the same artist? I suppose it could've been, now that you bring it up. Similar color scheme, same misty vagueness."

"Yes, it was," she said simply.

"I liked it. But I'm sure Horace appreciates it as well. And well it is he has it." He didn't look into her face as he spoke, but sipped slowly from his cup instead. When he did raise his eyes, he asked, "Do you know the artist?"

She answered indirectly. "I have several others upstairs. Would you like to see them?"

"You can be certain I do," he said, genuinely intrigued.

"So we shall, then, but after tea, of course. No need to be uncivilized." And she turned the talk abruptly to the weather, then teased the priest by pouring a second cup of tea. It took another ten minutes for her to petitely place her cup and saucer on the tray, arrange her napkin and spoon daintily beside them, and suggest that they could now go up to the second floor.

"We'll use the servants' stairs," she said, "but it will be somewhat tedious for you, I'm afraid." He didn't ask why, but the reason became apparent when she opened the door in the kitchen that hid the back staircase, literally the servants' stairs from the days when each society home had its own staff to do the family's bidding in return for living-in plus a monthly pittance. There in the narrow stairway was a chair lift, which Annie used to ride regally to the top. The priest had wondered how she negotiated the beautiful curved stairway to the second floor that graced the edge of the main foyer. It seemed too strenuous a trip for an elderly woman, even if but once daily. Now he knew how she did it.

He now wondered how long the apparatus had been there. Like the ceiling fan, it flaunted tradition and was an addition to her old home not easily acknowledged by older aristocrats like her. For that reason the priest refrained from comment. He walked gingerly up the stairs behind the slow moving contraption, in silence, because his hostess had chosen not to speak either.

The lift halted, and he waited for her to step out and fold the seat upward so he could step past. As he waited, he reflected upon the fact that he'd never been above the mansion's first floor. Not in all his twenty-eight years in Algoma! As proud as she was of her home, and as many times as she had taken him through its first floor, proudly pointing out antiques and historical oddities, Annie had never offered to show any more of it to him -- or anyone else he knew of, for that matter. *Probably because she didn't want to walk the stairs or give away the existence of the lift! Or maybe that sort of thing was just not done -- you know: show the bedroom floor to guests. Interesting that she would now!* Perhaps that would soon become clear. He waited patiently for the gallery tour.

"You haven't seen this floor before, have you Father Pastor?" she said with a precise formality. "Actually, it's rather

99

mundane; just bedrooms. The real treat is atop the third floor, above the attic: the cupola. It's the highest point in a private residence in Algoma. When I was a girl we three sisters used to climb up there and, with a spyglass, pretend we were pirates. We could see clear to Burger, which we pretended was an enemy port because you could see the river on this edge of Burger, and also occasionally there might be a small boat rowing up and down on it. But we're not going up there today," she sighed.

She stopped abruptly outside a bedroom and pointed inside to one wall. "Here's one of the other paintings," she said simply, and stepped back to give him a full, unobstructed view.

It was smaller than the three others he'd seen, but unmistakably from the same brush. A pastoral scene like the two at the Library, but almost all forest trees; not the meadows he had seen two days before. It was bright, nonetheless, and very appealing. He said as much.

"You seem to like everything of this artist's," was all she said. "Here's one you may not like," and she stepped into the room, to point out a larger watercolor just down the wall from the pastoral scene. Bright like the others, this was less dreamy: a well-captured single rose in a delicate glass vase.

"Ah, but I do like it, Miss Annie," the priest said. "Why ever should I not?"

"It's not an oil," was all she said, as she turned and silently indicated the other three walls, each with a framed piece; and not one like its neighbor.

"Are they by the same artist?" he asked, as he realized their difference from the others.

"Yes, they are," she said, and allowed him to silently inspect each more closely. She appeared to be awaiting his appraisal, which was slow in coming, as he took time before each piece. One was a charcoal of a cabin interior. *Lincoln studying by the fire, sans Lincoln* might be an improbable title. The next was a pencil drawing of flowers randomly strewn on a tabletop. And the last appeared to be an ink sketch -- exquisitely detailed, not at all vague or misty -- of a fine old home, not unlike the mansion, with stately trees beside it and the rear of a horse carriage peeking round the back corner of the huge porch.

"If these are all by the same hand -- and I believe you when you say so -- they're marvelous, and show a broad range of talent. Are they all from roughly the same period in the artist's life?" She didn't answer, but turned in the doorway instead, walked out to the hall and stepped into a second bedroom.

"Here, look at some others," was all she said.

He followed obediently. These four were portraits. One – of a man -- was rather large; the other three were women, one of whom was older. And all were in the same dreamy style as the initial three, but more detailed and not so brightly colored, and the people, though well rendered, were all notably sad. He detected facial resemblances. "Are these portraits of actual people?"

She nodded in the affirmative.

"Then, relatives of each other, I suppose."

Again the same nod. "Members of the artist's family," she said.

"Oh, really. How do you know?"

"Because," she said simply, but with dramatic effect, "I painted them."

He must have looked awkwardly funny. He certainly felt that way. All he could manage was a peculiar tiny laugh, which he followed with: "You can't be serious? Annie, these are marvelous. I had no idea ..."

"Few people do, I dare say," she said, and stepped back into the hallway.

He fairly shouted: "Annie, please don't walk away. These are marvelous."

But she called to him lightly: "Don't you want to see the others?"

She had disappeared into yet another, smaller, room, and when he caught up with her, he could see it might be her studio. Or maybe simply a storeroom -- he couldn't imagine anyone painting in a room with what seemed so little light. Numerous canvases were stacked about: on the floor, on several tables, against two of the walls. What was evident to his quick survey of the room was the uneven quality of the pieces visible to him. He guessed they might comprise a lifetime's output, and further that they comprised both finished and unfinished works, including obvious early attempts.

She explained this was her inner sanctum, her studio indeed; and that he was allowed merely a glance, because an artist doesn't show things until the vision is completely rendered. "So. You've seen them," she said with a smile, and stepped into the spacious central hallway, quickly closing the door behind her, having allowed really only the briefest of views into that room.

He was dumbfounded, and she was enjoying every precious minute of it. He touched her on the arm, as much to stop her from walking away as to regain the conversational momentum. "A moment, please," he said. "This is astounding. Don't rush away without my being able to ask about it all."

She smiled sweetly -- too sweetly: "We can talk, if you wish. But downstairs, where it's cooler." As was, he noticed for the first time, indeed the case. The upstairs was quite warm and stuffier than down below. Then he realized what in theory he knew already: that a good deal of the downstairs warmth was by this time of day up there with them, and with the second floor ceilings not as high as the lower level's, the upstairs air, unable to rise as high as downstairs, was trapped closer to the floor and affected them more. And the upstairs windows were all, as far he could determine, shut tight, besides. *How can anyone sleep without opening every window to get a breeze through here?* Then he remembered that Annie had no air-conditioning. *Perhaps her nephews have a point; breathing must be more difficult up here.*

"But, please, you must allow me to see these a mite longer; especially the portraits. We rushed out of that room, I'm afraid."

"Certainly. But will you mind if I wait downstairs?" she said, turning toward the stairs. Then she obviously thought better of it. "On second thought, it would be impolite to abandon you. I can remain awhile longer," she said coyly.

He quickly guessed she didn't want him wandering about unchaperoned, to perhaps enter her studio again. Only after he returned to Saint Helena's did he realize she might not have wanted him in the other two rooms either, one of which was surely her bedroom, where the oxygen tanks were.

"Thank you," he said gratefully. "I promise not to give either of us heat stroke." And he disappeared into the portrait room. Upon reappearing moments later, he quickly stepped in front of first one piece and then another that she hadn't pointed out, both in the hall. "Thank you, again" he said when he had finished, indicating that they could now descend to the ground floor.

"After you," she said. "It will be easier for you to go first." *And safer, too,* he thought. *No danger of a quick peek at any forbidden chambers.*

In the kitchen, she asked: "More tea?"

"Yes, I believe so." *We have so much to discuss!*

"I'll brew some. This has most assuredly lost its flavor," she said as she brought the now tepid teapot out from under its cozy on the sideboard.

He went onto the cooler summer porch as she refilled the kettle, put it on the stove and puttered about with the rest of the tea ritual. He waited for her to join him before speaking again, the better also to regroup his thoughts. While they waited for the kettle to sing, he said. "You amaze me. Where on earth did you get such wonderful talent? And why ever are you so bashful about it?"

"Few people have ever called me bashful. Aren't you a dear!"

But he wouldn't be put off. "Now, dear," he said, deliberately using her own word, "you know that's accurate. You haven't told a soul, I dare say. Why in heaven's name hide your light beneath a bushel?" *There must be an explanation, but I can't for the life of me imagine what!*

"It's not the thing a lady should do, you know."

"Paint?" he asked, incredulously.

"*Tell people,*" she replied. "One does not blow one's own horn. It's unseemly. Particularly for a lady."

"Annie, these are good enough to sell. Those I've seen certainly are. The two at the Library fetched quite a price. Why on earth not offer them to a dealer?"

"Oh, but I do, Father John."

She has no end of trumps up her sleeve. Did I call her bashful? "You do? I mean to say, I'm pleased; but amazed nonetheless. Have you sold many?"

"Ten or twelve."

"That's not many for as ol ... " He coughed to cover his embarrassment, and corrected himself: " ... for as long as you've obviously been painting."

"Oh, it's just recently that I've attempted to sell any. The last year or so."

"A year or so!" he said in awe. "It's a pity you waited so long,"

"I didn't think anyone would be interested. But when the thought occurred to me, I offered one small oil. I was surprised there was a buyer, but absolutely astonished at the price he offered. I never dreamed that could happen; never in my wildest fantasies, I assure you."

"Do you sell them locally? I wouldn't know how to do that here. There isn't a dealer or agent that I know of; not outside Saint Louis or Chicago."

"It's in Saint Louis! A ... friend ... takes them to a dealer for me. I must say, it's wonderfully flattering to have someone express interest in something you've done -- and be willing to offer money, besides."

"Well, you certainly have others that can fetch a fine price. I mean, just the ones I saw so briefly upstairs could."

"Oh, those are too precious to me. Most of them, anyhow."

"What about those in your studio? There must be a hundred or more ... "

"Oh, those. They're hardly ready." It sounded like she had just dismissed an expensive ball gown as *that old thing*.

"Surely *some* of them must be." the priest said encouragingly.

"One or two at most," and she paused. "Maybe."

"Well, I for one hope you find more that *are* ready; or *get* some ready. You could be the next Grandma Moses." To his mind's ear that sounded peculiar, so he added: "Would I offend if I said I'd like one; particularly before the art world drives the price up too high for me?" He smiled, but meant every bit of it.

"You're just flattering a vulnerable old lady," she said, lightening the conversation with surprising dexterity. "Something only a cad would do. I know your type, sir." And she smiled.

"Trying to boost the price already?" he accused, in jest. "I *am* serious," he added after a short pause.

"You can't be, of course."

"But I am. The piece in the hall, the one you did not mention -- you know, the last one I looked at: the tree by the brook, and the little waterfall ... "

"Well, as an artist -- not to mention, a woman -- I'm entitled to some dalliance, am I not? I'll have to think about it," she said firmly. He couldn't tell if she were truly serious, playing serious, or seriously playing with him.

"Are you trifling with me, Miss Verden?" he said playfully.

"Certainly not, Father." He still couldn't tell.

"What might you want for it?"

"It's priceless." Now he was sure it was a game. *And two can play!*

"Priceless! Then I am lost." And he rolled his eyes.

"When an artist won't sell, the piece becomes priceless."

"Life is hopeless," he said, his hand melodramatically over his heart.

A smile finally broke across her face. "I really do prefer to keep it. But I promise to think it over. But that's all I'll say on the subject today," she said, as the kettle in the kitchen started its song and she rose to attend to it.

When she reappeared with the tea tray, she frowned as though having forgotten something, set the tray on the table with the request that he begin to pour, and turned back to the kitchen. She reappeared with jam and a small plate of biscuits, and noted with approval the priest's beaming countenance.

They sat a long while, sipping contentedly, before he broke the silence. "If Horace told you I returned the painting, he must surely have also said why."

"Yes," she said simply.

"Is it indelicate to ask why you gave him the money I found bulging out from behind the painting?"

"Frankly, yes. But I'll tell you anyway," she said, turning slowly to face him, smiling ever so slightly.

"The poor man did me a favor," she said slowly, as though choosing the words carefully, "and I simply wanted to thank him. I didn't want the money to fall into the wrong hands, as it might well have if I'd left the envelope beside the package on the steps. That's where I leave everything he's to pick up."

It could have been put inside a package with the painting, instead of inside the frame!

"And I didn't want to embarrass the poor man. After all, he may have thought the favor trivial, unworthy of such recompense." *Poor man! Why call him that -- twice? To embroider the story? Or does she really pity him?*

Father John didn't know how to take that *or* respond, so he decided to accept it at face value -- or at least, to hide his doubts. "Well, I'm glad he could reclaim what was rightly his. I'm sure he's enjoying the painting as much I did."

The look on Annie's face was peculiar and puzzling to the priest. After a moment, she nodded as if in agreement, and rose to begin putting things away. Apparently their soiree was over. He rose too and helped carry things, over her mild objections. Responding to her less than subtle message, he said: "Thanks, Annie, for a lovely afternoon. The tea and biscuits were a fine dessert to the marvelous feast of your lovely paintings." *You elegant old codger!*

"You're most welcome, Father. But please be so kind as to be discreet about my art. I am ... a lady. Not *bashful!* And I prefer to bestow my charms upon those I choose. I'm sure you understand."

He both did and did not -- understand. But he nodded appreciatively and took his leave through the back screen door in silence. The walk back home was slow and thoughtful.

CHAPTER XIII

Life in Algoma, Father John had begun to fear, was returning to the depths ordinarily reserved for the early weeks of summer: *dull, duller, dullest.* Were it not for the nephews' near paranoid concern for Annie's health, a fear for which he could still find no firmer footing than the second floor heat, *and* also for Annie's peculiar interest in Horace, the priest might just have to take home some thriller from the library to satisfy his craving for a mystery. Annie's artistry and Horace's Algoma roots were certainly interesting news; but they were tucked firmly away in his heart, where, he decided, they should remain.

Besides, they *were* answers. And to mysteries, but not like Father Brown's. He loved Chesterton's unlikely hero, secretly envying his ability to stumble again and again into just the right sort of mysteries, those that allowed for people-patching as well as crime solving. Another Brown, Maisie, was to restore his original quest, or at least administer a mild splash of adrenaline. Later that week at the bank she said Annie was indeed using her oxygen.

"How do you know, Maisie?"

"Cuz, rev'ren, she done finished one tank."

"That's marvelous, Maisie. That means she wants to help herself. I'm so relieved," he said. Maisie beamed.

"Bout hepin at your rec'ry, rev'ren… I'm not sure I be able. Not yet, no how. I only got jes one day off and I kinda got used of my freedom."

"Not to worry, Maisie. I can get by this summer. It's the fall I'm thinking of. We can discuss that later."

"Thank you, rev'ren, sir," she said. "But I ain't forgot."

"Nor have I, Maisie. But there *is* one other thing. Do you know *when* Miss Annie uses her oxygen?"

"At night, I speck, rev'ren."

Of course! Doubtful Annie goes up and down those stairs much, even on that lift. And it is *hot up there; even if she opens windows, which I don't think she does; not all of them, anyway. She'd hardly go up there in the day's heat.* "You don't know if she uses a fan at night? It's got to get warm in that upstairs."

"I don't fer sure. But those boys bought her a fan, member? I unplug it when I clean, and it's always plug back in next time I come round. You be right 'bout those rooms. They sure hot. Daytime, anyway, when I come round. Don't speck they be much better most nights, hot as this summer been."

"Well, *plugged in* is a good sign. You mentioned *rooms* upstairs. Just how many are there?" He already knew, of course.

"Six, rev'ren; and the bath. It be a big house."

"Goodness! Six bedrooms!"

"Well, five fer sure be. I don't know 'bout the other one, cause she don't never let me go in there. Know what, though? When the boys were round, I had a couple extra rooms to clean, a course. But before, and since they done gone too, ever so often I got another extra one to clean. Can't figure it none, why Miss Annie sometime sleep there."

"The same extra room each time?"

"Yeah. But still don't make no sense."

"Well, every so often she just may not be able to sleep in her own room." But he didn't for a moment believe that.

"Say!" he said, as much to mask that disbelief by changing the subject as to inquire about something else he just remembered and genuinely wanted to know, "How does she use that oxygen, anyway? Does she wear a mask?"

"She use a what-cha-ma-call-it hooks on her ears and sticks in her nose on a tube comin outta the tank. Purt near sure she does that all the time."

"That's pretty much the way it's done in hospitals. Sounds pretty easy, not? Or would you know?"

"It sure be, cause she have me hep her onced. Jes after she got it. I think to make sure she knew how herself."

"That was a good idea," the priest smiled. "You don't help her with it any more, I suppose?"

"Well, only if she be feelin a little puny some. Ever so often she go to bed early. I don't think it's no nap, cause it's too late of an afternoon. But if I still be there and she be feelin like that, I hep her. Not too often, though."

"Well, Maisie, our collaborating is working out just fine. Thanks. I'm glad you have such good news about Miss Annie's health."

"Oh, I be glad clab'ratin. You can count on me, rev'ren," she beamed.

Indeed I can. And happy for it, certainly!

Back at Saint Helena's he realized he didn't think to ask if Maisie opened any windows for Annie when she put her to bed those occasional afternoons. And he couldn't get Maisie's other revelation out of his head. *Who in heaven's name uses that other room? Surely not Annie! Who, then? Maybe Annie really does have some breathing problems. How can I ever become certain about that? Because, if she does, maybe I've seriously misjudged those nephews!*

Much as he hated to gainsay his intuition, he had to admit the pendulum was swinging back in favor of the two young men. *Maybe they really are as nice as they seem, and simply care about their elderly aunt after all. You suspicious old celibate! I wonder how many other folks are as devious-minded as you!*

The priest looked up suddenly and narrowly avoided bumping headlong into Lumpy Wurtz. "Oh, pardon me, Richard. Wasn't watching where I was going. Making a deposit?"

Richard Wurtz was a taciturn farmer, a bachelor from a large farm family of German stock, all the rest of whom had long since married and left the farm. Tall and wiry, Richard had worked the farm and lived on it with his parents, caring for them till their deaths, then staying on the land afterwards alone. It was the only life he knew, and one he loved dearly. But you wouldn't know that from anything he normally said, because he was a man of not merely few words, he was a man of almost no words -- for most folks, except Father John.

Lumpy -- his boyhood moniker bestowed by an older brother after his fall from a hayloft gave him a bump on his head for a few days -- had suffered a more recent fall from his farmhouse roof one winter day several years back. During the down time when farmers have little to do in the fields and spend their hours catching up on repairs, he was caulking an upstairs window and lost his footing on his steeply pitched roof. The broken arm, cuts and bruises weren't that serious, but his concussion was. His hired hand probably saved his life by fetching Feldspar's and getting him promptly to the Burger hospital.

Father John came as soon as he was called that gray, mid-December afternoon and anointed the man, whom he found mumbling a mile a minute. *Coma talk* was his name for it when he told Richard about it later. But it was really morphine. The priest was so touched by the man's sad appearance, and so flabbergasted to hear all those words, however occasionally indistinct, come tumbling out of this supposedly speechless mouth, that he spent the night with him, holding his hand, fetching water, and once calling the nurse to administer another sedative when his patient seemed particularly distressed. That night Father John came to deeply appreciate this erstwhile quiet man whom most people dismissed with the mention of his nickname, as if to intimate that a childhood mishap had defined him forever as someone unable to think, let alone converse, properly.

He heard the soul of a poet come pouring out that night as Richard spoke of sharp, crystal mornings when the only thing moving for as far as he could see was his own breath tiptoeing out of tingling lips. He spoke of the overpowering smell of July corn, and of the sounds of the creek that wandered his property, watering windbreak and small game alike. The priest learned the names of each of the cows in Richard's herd, and discovered that he'd kept two horses from the plowing days of his youth before his daddy bought a tractor; kept them long after their usefulness had faded; kept them till they died, because he hadn't the heart to put them down. And after that long, strange night the priest also knew that Richard had flowers in his small garden as well as the tomatoes and peppers he sometimes sold in town or gave to friends. The man simply wouldn't stop talking. And Father John couldn't stop listening.

He learned that the cattle's watering trough held more than ten bathtubs could; and that blue jays are pretty, but rob other nests; and that Richard for that reason didn't scruple to shoot them. He heard him say he liked being up before the sun so he could get a jump on the morning's work and also watch the world wake up. And he heard him describe the pattern of the farm's arousal from sleep: the chickens spoke first, but the cattle were already awaiting him, no matter how early he arrived, stamping softly in their stalls, moaning a greeting when he walked in, thanking him with their tails as he poured their mangers full of morning grain. It was plain from his mumbling that he conversed with his herd. And he believed they responded.

He spoke glowingly of the breezes in the farmyard trees, and of the birds that called those trees home. The blue mists of morning were not preferable to the gray of dusk; they were just different experiences. And unlike townsfolk, Richard liked fall most of all -- not spring -- because of the proud relief he felt when the harvest was in, and a chill touched the air, and nature began to bed down for a three or four month nap, and Richard could breathe deep and proud and contented as he roamed his quiet acreage, not the master of all he surveyed, but its friend.

These things and more he heard that night. These insights and others the priest received into the soul of his bachelor cat-licker, this tall, quiet man. And these feelings and many, many more endeared the man to his pastor; and the pastor to the man when, later, Father John could tell him how moved he had been that long night which ended too soon with the call to morning Mass. Thereafter, he was a regular guest, usually in spring and fall, at the farmhouse with the whispering trees and talking herd, where he sat on the porch and conversed as much with the land as with his host, and drank fresh, sweet milk in endless toasts to the cows it sometimes took all afternoon to name. And the bond between the two had long since deepened in silences as well as quiet conversations, as Father John Wintermann came slowly to learn that not all poetry is spoken aloud.

"Yep, a deposit," the man said, simply, and winked. "Heck fire, Father, I'll not be payin fer any loans."

They both smiled, and Father John asked if he'd been to town lately now that his wheat was in, or if he'd only just come in today.

He had come in lately, indeed. *To the Libary thang.*

"Oh, the auction? I didn't see you there."

"Saw you. You was busy looking at those paintings. The ones the out-of-town ladies bought."

"Why, I was, all right. Too expensive for me, sad to say."

"Shoulda bought a small one."

"There weren't any, Richard. No smaller ones; just those two."

"Not there. Before. I got one before."

"You bought a painting by that artist? Before the Library Sale"

"Sure did."

"By the same artist? Are you sure?"

"Yeah. Same funny squiggle at the bottom corner; s'posed to be a signature, I guess. Anybody write like that oughta be shamed of hisself, specially if he's that good a painter."

"Where'd you find one?"

"Asked Horace Denver if he could find me sump'n pretty I could 'ford. An he did. Brought me a little bitty pitcher of a pansy. Got it in the kitchen, where I see it ever day. Surprised to find it was a same as those other two at the libary."

"Well, weren't you lucky." *And me too!* "Got it cheap, did you?"

"Shoot fire, a whole lot less'n those two ladies paid out."

"Well, when the weather slacks off, and your work too, I'll be out to the farm. I'd like to see that painting."

"You can bank on me bein there. Just you come on out," he said, and moved away to a teller whose window had just become available.

"I will, Richard," he called after him. Always Richard! Never Lumpy! Not since that all-night corporal work of mercy at the Burger Hospital. Never, never, never again *Lumpy*.

So, Horace is the art connection. Well, I'll be! Wonder who got the Last Supper?

CHAPTER XIV

"Yes, this is Father Wintermann." It was long distance from Pittsburgh one week later, just after August had begun with no relief from the heat but with the certain promise of record receipts for the beer stand at next weekend's Lutheran Bazaar. Anthony, Annie's one nephew, was inquiring about his aunt.

"Hello, Father John. I don't wish to seem an alarmist, Father, but is my aunt all right?"

"As far as I know, why? Have you tried reaching her and got no answer?"

"No, nothing like that. Remember our concern, Joe's and mine, about her bronchial condition? Well, we took pains to get her some oxygen before we left and we're wondering if she's using it. We haven't heard anything about that and, frankly, with a woman like our aunt, it's a little ticklish to just come out and ask. I was wondering if you knew anything. Like, does she go out at all? Is her breathing okay? And would you know if she's using that equipment?"

He almost told them everything, but decided not to at the last moment. Among other things, he didn't want them to know how many pains he'd taken to check on Annie and their story as well. While he was personally relieved Annie was doing something sensibly therapeutic, *they* could wait awhile to learn of it.

"It's been much too hot for your aunt to go traipsing around," he stalled. "far as I know, she's been very sensibly staying indoors. Except for the occasional evening in the swing on her porch. I'm sure you understand how important that is to her. I mean, the fresh air, the company and all."

"Yes, evenings on the porch! She couldn't do without them! Have you visited since we left?"

"Yes, several times. She seems fine."

"Her breathing's not labored? No coughing, no gasping for breath?"

"Nothing like that. Is that what you noticed here?"

"Traces of that, yes," he said. "Excuse me," he said, abruptly. The priest could hear his brother in the background. When he came back on the line, he said, "Joe said to say hello. We're both appreciative you keep a close eye on her. With her age and all, we worry more and more about her. You wouldn't know if she's using the oxygen, would you? We got her some, as I said."

"She hasn't said anything to me about it," he responded truthfully enough. "But like yourselves, I'm not sure know how to go about finding out, either; not without just up and asking. With someone like your aunt, personal issues are taboo, I don't need to tell you. Unless she flat out says so to me, I'm not sure how I could know for certain. I suggest *you* ask her. You ought to have better luck than myself. After all, you got her the stuff; it shouldn't be out of place for you to inquire about it, should it?" *Ball's in your court, fellas!*

"There is one way, Father."

They're determined!

"Do you know the black girl who works for her? Maisie? We asked her to encourage our aunt to use the oxygen. Is there any way you could find something out from her? We're not sure we have her number. Or, more precisely, we can't get her to return calls. And we're wondering if there's a better number or something. It's a bit confusing, but the bottom line is we feel stymied. We thought we were so clever enlisting the girl's aid, but now we can't contact her. Perhaps you can, and either have her call or, if all else fails, call us yourself. Either of you can call collect. I'll gladly give you our number here."

That he could do, of course. He promised to call back soon. And then the young man just hung up, without any more pleasantries, without even so much as a comment about the dreadful weather. Father John determined he'd give them reassuring news in a week or so, but no sooner. He'd see Maisie at least once more by then, though he didn't expect to learn anything new. They'd be happy with that *and* also reassured they hadn't spent their money foolishly.

Money! I'm still not sure that isn't what they're really concerned about. But, how to be sure about that? No real answers to that either. He was still bumping into a stone wall.

His weekend was occupied with usual things: Masses, plus a wedding between a sweet young thing and her equally vacuous beau. More and more he brought himself to such sacramental moments less and less eagerly. He found his faith in the younger generation these days more often than not *stretched*, to put a nice face on it. There seemed to be more *children* marrying. Regardless of chronological age, many of these couples looked immature, acted immature -- *were,* no doubt in his mind, immature. Less thoughtful, altogether! *I hope it isn't just old age, this creeping cynicism. But then, it isn't just me; there are lots of other priests equally discouraged about this sort of thing.*

A lone ray of hope had been supplied by the Carbondale Newman Chaplain: the marriage age was rising among his college set. And, in his estimation, there was also noticeably more maturity in that crowd with regard to things like choice of partners, reasons for marrying, and approaches to child-rearing. He sincerely hoped it was a trend, but his own experience didn't corroborate that, though, he realized, many of his marriages at Saint Helena's involved people who -- one or often both -- hadn't finished college. Some even had no higher education at all, and were lucky, for that matter, to have finished high school. More to the point, they didn't seem all that experienced at life. Many were chronologically very young to boot. Then too, he knew very few of them, what with one of each couple ordinarily coming from out of town, and the Algoman, like as not, no longer living at home either: in effect, also a stranger.

The following Tuesday he decided to walk that marriage license over to Dixie instead of mailing it. And after a chat with the charming County Clerk, he was about to leave the courthouse when Pat Kelly came up to say hello. In his early fifties and a native of the county, Pat moved into Algoma shortly after law school, to settle in his wife's hometown and set up practice near the courthouse where virtually all his cases would be tried. He'd made a successful law practice in the intervening years, doing general legal work. *A little of this, a little of that,* as he consistently described it.

"Father, what brings you out in this heat? A real sizzler, yes-sirree Bob!"

"Got that right, Pat. Just put the final touch on one of the biggest, hottest, longest weddings in years. Dropped off the license! Eleven couples in the party! I knew I was in trouble when we had to practice the processional three times. Three times! And the next day wasn't much better. Thought we'd never get out of there! And you know how hot Saturday was! Felt sorry for the ladies. Thought they'd surely wither. But youth's wonderful! None of them showed any signs of heat stroke. I was the only one feeling anything. And I can tell you I certainly felt it! Speaking of weather, I notice by your still starched collar that *you're* staying in the cool confines of this legal palace today."

"My momma didn't raise no dummies, Father." They both chuckled. "But glad to bump into you. Maybe you can give me some advice. Other day I received the oddest request long distance -- and then an overnight Fed Ex -- seeking a change to someone's will."

"That so unusual?"

"Well, it is if you consider that the will belongs to someone other than the persons making the request."

"Wondering why they didn't take care of it where they live? Or what?"

"That's part of it. But also, the person whose will it is lives here in town, and the requesters -- plural -- do not."

"Does sound odd. But why ask me? Ask another barrister. Plenty of them *here* any given day! What do I know from legal stuff?"

"It's not exactly a *legal* problem. You see ... " He stopped mid-sentence, looked around, then added: "Could we duck into an empty court room? There's one upstairs, if you don't mind. It'll only take a moment -- I hope."

The priest shrugged, and they both climbed the broad ornate staircase -- Father John a little slower than his companion -- to the second floor, where they found a courtroom with the privacy the lawyer wanted. When he'd closed the door and determined they were indeed alone, he turned to Father John to resume the conversation.

"All the papers are in order; the signature's notarized; the whole nine yards. And I'm the right guy, because I *am* her attorney. But if this *is* her idea, I just can't figure out why the papers originated in Pennsylvania, and why she didn't bring me in on it in the first place."

Pennsylvania raised a yellow flag. Father John could almost guess whose will it was. But he waited more or less patiently for Pat to say, if indeed he intended to. As nonchalantly as he could, he asked: "Where were they notarized?"

"Why? Not sure I thought to look, and I certainly don't remember. Here, I assume; where my client resides."

"That's why I asked. If they were notarized here, why were they mailed? And why indeed weren't you in on it from the beginning? But if they were notarized elsewhere, they must have involved a lawyer, no? So, just talk to that man. Or woman. Whatever."

"They weren't prepared by a lawyer. No indication of it: no legal stationery or lawyer's signature -- nothing like that. But they're in perfect order. It's a neat, signed codicil, in effect nullifying the rest of the will. I guess that's another thing troubling me. Why not write a different will? And why not retain *me* to write it? It's the standard thing. Codicils should emend, not nullify. And another thing, everything is transferred -- the entire inheritance -- to the two who sent the papers. Doesn't smell good."

"I think I agree, as you put the facts to me. But again, why ask me? Just ask another lawyer. Or, better still, your client himself. Herself!" he corrected himself. "I think you said *she -- or her.*"

"Yes. *She.* But that's why I thought of you. You see, you know her. Rather well, I believe. If this is her end run around me, I'd just as soon not have to speak with her. But if it's their end run, I'd *better* speak with her. I was hoping you'd know something that could help me decide. My client's Annie Verden. Has she said anything lately, or is there anything you're privy to, about changing her will or altering her intentions relative to disposition of property after her death? Something, anything, you know -- that you're allowed to speak about?"

Father John's intuition was vindicated after all, but he didn't feel good about it. "No," he said truthfully. "In fact, the only thing she mentioned lately even remotely in that ball park leads me to conclude she is *not* considering changing anything. But! You know Annie! Lots of things are possible!" *Though I don't believe cutting her nephews in on the estate is on that short list!*

"Well, isn't this a fine kettle of fish! I've got an apparently legal document which can change everything -- if it stands up in court. As, indeed, it could!"

"It ought to be easy to verify. Call her. If it's on the up-and-up, she'll say so. And that'll be that! If it isn't, you can smooth it out. Nicht wahr?"

"Well, yes. And no. I don't know your exact relationship to her, Father. Maybe it's different for you as a priest. But I'm a hired ... *servant.* That's the word, all right. Just like the old days! You know: when families like hers had servants -- whom they barely tolerated. That's the feeling I get. And I don't relish feeling that any oftener than I have to. I suppose I'm going to have to talk with the lady; but I'd sure like to get around it."

"Oh, come on, Pat. You guys command a hefty recompense. Just wade in there, do your job and collect your fee. That ought to be the Balm of Gilead for your troubled soul." The priest was smiling. But the lawyer wasn't.

"You may not believe this, but Annie's a pro bono client. I was led to believe she can't really afford much at all. Didn't want to believe that, but got it on good authority." He paused. "I don't charge a thing."

"You can't be serious? Who gave you that information?"

"Bob Lanner. Called me, dead serious, shortly after Annie wanted to engage me fifteen, twenty years ago. He should know; he's her banker. And, far as I know, nothing's changed about her finances. *Or* our arrangement."

"Yes, he *would* be her banker. Who else! And I suppose he should know. But, tell you what: Mr. Lanner owes me a favor. I'll talk to him before you get hold of Annie. In any event, my considered opinion is that you need to talk with her. And the sooner the better! So I'll get back soon as I can. I'll go over to the bank right now, in fact. If I'm lucky, we settle this today. Tomorrow at worst! *Amicably* would be nice; but settle it one way or the other. To repeat: I don't think she intends to change her will. Could be wrong. But I don't think so."

"Glad I saw you, Father. I'll be waiting." The two shook hands before stepping into the hallway and descending the wide staircase. Then Father John turned and walked out the front door of the courthouse directly across to the First National Bank of Algoma.

CHAPTER XV

John Henry Wintermann was a peculiar blend of modesty and passion, and an enigma to people who tried to categorize him. He bent his energy to the tasks of vision without an iota of self-aggrandizement. The only sense the uninitiated could make of this single-mindedness was to call it self-confidence: a totally erroneous reading. Quite the contrary! It never occurred to him to think himself important, let alone confident. These had always been farthest from his priestly mind. Rather it was more like tunnel-vision: always it was the issue at hand, the vision to be served, the task to be completed that deserved and got full attention. Nothing more. Or less. Such was his mind as he walked into the bank that Tuesday morning, a man on a single-minded quest. He headed directly to the office of Mr. Robert Lanner at the rear of the bank.

He had forgotten all about that being his day to see Maisie. When she hailed him from across the lobby he was startled to be sidetracked. He looked around in momentary confusion to search out the voice, which, jarred out of intense concentration, he was unable to readily identify. When he caught sight of Maisie's smiling face, he broke into a shy, self-mocking grin.

"Pardon me, Maisie. I was in such a hurry I didn't even notice you. How are you?" he said, as he walked the several steps across the lobby to her. She was standing by a potted plant at a large window whose sill she was dusting.

"Oh, I be fine today, rev'ren. Longs I stay cool in here. And I be fixin to keep doin that, fer sure," she said, and flashed a huge grin.

"Don't blame you for being indoors, Maisie. It's a potboiler! Probably get over a hundred again." Once they were close enough and could lower their voices, he asked: "Anything new on Miss Annie's oxygen?"

"Nothin. 'Cep, maybe I forgot to say she told me she wants to make it come out slower, bein's how she don't think she need all that much, and it so 'spensive and all. I think she be worryin 'bout who gonna pay fer a new tank when she be needin one."

"Has she finished the second one, then?"

"Oh, no, sir. I think she jes be lookin ahead. But I don't know nothin 'bout that stuff. An I told her. She seem to take it all right ."

"So what will she do?"

"Don't know fer sure. But, she say she be askin Horace Denver. He might could do sump'n. Lease wise, she thinks so."

"Might just have a point there, Maisie. He *could* know something about it. Or figure it out, all right. Is Miss Annie fine, otherwise? As far as you can tell?"

"Oh, yes, sir. Sassy as ever." And, winking at the priest, she giggled.

"I know what you mean, Maisie. Thanks. I'll be in touch."

"Bye, rev'ren," she said and went back to dusting, still giggling.

As he turned again toward the president's office, he caught sight of Roberta Sue Langley, Mr. Lanner's private secretary, whom everyone called Bobbi Sue. If you knew her *very* well, he'd been told, you could call her *B. S.* "Bobbi Sue!" he called pleasantly.

The well-dressed and very proper middle-aged lady -- Algoma's version of a classy businesswoman -- turned and broke into a broad smile. "Well, hello, Father Wintermann. How's my favorite chef?"

It was her usual greeting and a reference to their first meeting. Shortly after his transfer to Saint Helena's, the new pastor was punctiliously performing his self-appointed duties as chief hamburger flipper at Saint Helena's Parish Picnic, a major fund raiser typical of small parishes and which, not many years thereafter, parishioners decided to discontinue. Too many such shindigs had cropped up around the diocese and a fair number within Algoma itself -- the Lions and Lutherans each had one, the VFW and American Legion held street fairs, and the county fair went on nearly a week. The bloom was off that rose. Competition was too fierce and self-defeating.

So they switched to an early spring auction instead. Over the years it grew into quite the social event. Staged at a time when little else competed -- late-March, when they could easily avoid high school basketball, the only thing to possibly challenge attendance -- it never failed to bring out most of the townspeople. And over the years the parishioners became quite inventive as to auction items: trips to resorts, people who promised to work for others -- *yard-,* etc., *slaves,* in auctioneering jargon -- and Father John's specialty, a meal for four or six, prepared and served at the rectory. A dress-up affair with a different pre-announced menu each year and several appropriate wines, it had become the evening's highlight and sparked spirited bidding between rival couples, the winners often inviting the losers to the meal. And though Father John actually did prepare and serve it -- and dine with them also -- he privately admitted that his friend, a Saint Louis chef, advised him and even did prep work on occasion.

In fact, several years after that first culinary encounter, Bobbi Sue and husband actually attended an auction meal, guests of the high bidders. But her fondest memory was from his first Algoma summer when the new and heavily perspiring pastor was burger flipping with panache at the Parish Picnic.

"Not very hungry, I'm afraid," he grinned. "The heat, you know."

"Isn't that the God's awful truth!" she drawled, the smile still firmly in place, "I swear: we're going to dry up and blow away! Where you going with such a determined look on your face? The restroom's over there," she said, hoping for at least a tiny blush.

"Naughty, naughty," was what she got instead. "I'm here to see your boss, actually. Only, he doesn't know I'm coming. I just got a revelation, sort of. You know, a bright idea. Is he in?"

"Where else would he be? Of course, he's in. And I don't think he's busy, but let me check. Have a seat ... " They were by then directly outside the inner sanctum, and as he sat down she went into her own office to pick up the phone. In a moment she poked her head out and said sweetly, "Go on in. *Now* he's expecting you."

"Thanks, Bobbi Sue," he said, and entered the comfortable office.

"Mornin, Father John. How're you today?" he said with a smile. *Still the charmer. Even after our last conversation. Amazing!*

"As well as an aging cleric can be in this God awful weather. Yourself, Mr. Lanner? Your diabetes ok?"

"Fine, thank you; and the diabetes ain't actin up any more. Have a seat. Yes, it is fee-rocious, and the humidity makes it absolutely unbearable. What you up to today, Father John?"

"Well, actually, I'm asking a favor, one I'm not sure you can grant, or even want to. So I'm a little hesitant. But I'm here, nevertheless. In a nutshell, it concerns one of your customers' finances. Now, if I'm overstepping the bounds of propriety here; if I'm asking you to divulge things you shouldn't ethically speak about; or even if I'm just meddling in things friendship has sealed your lips to, I'll understand if you refuse. I want that known up front, okay?"

"Understood, Father. What is it?"

"It has to do with Annie Verden's financial situation."

"Specifically what, Father?"

"I just talked with Pat Kelly. He says he represents her for free ever since he took her on as a client. And he said he does so because you urged him to."

"That's essentially correct. Quite some years ago now, I explained her situation was ... delicate. That's the word I used. I didn't specifically ask him to do it for nothin, but I let him know Annie's not as flush as people believe."

"Was that the truth?"

"Yes, indeed."

"And is it still? The truth, I mean? She still ... financially strapped?"

"I do believe so -- not as bad as earlier. But, as I think I told you when last we spoke, I am not *precisely* sure of that now."

"Our last conversation?"

"Why, yes, Father. I distinctly 'member sayin as much."

"You never mentioned Annie, Mr. Lanner."

The banker looked perplexed momentarily, then troubled. "I thought I had. Perhaps I just assumed ... *Annie,* you see, is Horace's mama."

Now it was the priest's turn to look perplexed. And troubled. All at once. "I didn't think ... " was all he could mumble.

"Oh, yes, Father. Annie! *And* she hasn't near the money her family once had. Her daddy spent somethin terrible afore he died. Besides, the Depression was unkind to them, as it was to most folks. A course, Annie's pride has consistently concealed all that. She lives rather frugally -- has, all these years. It's why her house is so run down. I'd never let on 'bout anythin concerning that -- never *have.* Wouldn't divulge such things 'bout my worst enemy; but for Annie, why ..." He paused, silent some ten or twenty seconds. Father John continued to sit numbly, considering the implications of what he'd just heard.

The banker continued: "That's why I've never begrudged her any help. Even though, I do admit to a little pique now'n again, as I think I told you. But I'd never deny her anythin, really. Particularly since she's been generally reluctant to ask for very much a'tall, and rather pars'monious when she does."

"I must say, not only didn't I know before, but I'm dumbfounded to learn it now. And I'm worried about her. Is there something I should be doing? Or my parish? I imagine right off that would be awkward, but if she needs help ... "

"No, no, Father. She isn't *destitute*. But I had no scruples -- nor any now -- asking someone as successful as Mr. Kelly to forego a fee or two. And I must say, he's been the soul of discretion. At least till this moment. I wonder why he might ... *slip,* so to speak ... *now.* He *did* rely on *your* discretion, surely?"

"You can trust him -- as he trusted me. He had a dilemma, and I can advise him now, thanks to this conversation. Not that I wish to share Annie's connection to Horace. That secret's safe. Rather, I needed to verify that Annie's finances aren't as cushy as people think. For which information, thanks. But I must say, I'm still reeling under the news about Annie being Horace's mother."

"Yes, she surely is. Why else you think he got the surname Denver?"

"Well, I used to think his father's last name was Denver. After our recent discussion I assumed that it's because he came from Colorado."

"Oh, no. Not a'tall. It's a devious little play on words: Ver-den ... Den-ver!" He spoke with exaggerated slowness. "You see it, don't you?"

"Yes. Now! Whose idea was that?"

His tone of voice clearly indicated his lack of regard for it. "Annie's! I went along with it. Seemed harmless, and it was *so* important to her. But when he showed up -- at Annie's instigation, I don't doubt -- I got worried. That name's so much like hers! But no one figured it out. And after all these years, I think the secret's safe entirely." The two sat silently a minute or more, each lost in thought.

Then Bob Lanner spoke again. "You mentioned a dilemma -- Mr. Kelly's dilemma? Is it *me* now bein indiscreet in askin: 'bout what?"

"No, not really. But that's for Mr. Kelly to divulge. And, frankly, I don't think he'll want to. I believe he knows what must be done legally. And he'll do it. I trust him wholeheartedly."

"She's not in legal trouble, is she? You know I care about her. I wouldn't want anythin to hurt her."

"No, she isn't. And, yes, Mr. Lanner, I do realize your concern. Ever since we spoke, it's been abundantly clear how much you care for Horace's mother. To be honest, that gave me a very different -- better -- understanding of you. But, please, trust me! What Mr. Kelly needs to do -- *will* do, I know -- will be in Annie's best interest. And I repeat: she's in no legal trouble."

"I'm relieved. Cuz', 'member, if ever you learn Annie needs somethin, specially financial, I want you to come to me without a moment's hesitation."

"I promise. And thanks for this conversation. I'm in your debt."

"Don't mention it, Father," the older man said. And smiled. Sadly, Father John thought. But he smiled.

Walking back through the lobby, Father John greeted a patron or two and waved again to Maisie, then headed straight across the steamy street to the courthouse through the relentless heat, in search of Patrick Kelly.

Their conversation was brief. *Get hold of Annie. Pronto. Mr. Lanner admires your helping her -- so do I -- and thinks your arrangement should continue. Let me add that Annie's the one to settle the issue currently in your lap. She can say in a milli-second if this is to her liking. If so, you can be at peace; if not, there'll be hell to pay. But not with you!*

With mutual thanks and a handshake, each went his way: Pat Kelly directly to Annie, and Father John to his rectory, to think about his own relationship to the duchess. Things had definitely changed in regard to that. He had lots of praying to do.

CHAPTER XVI

Two days later, morning dawned already simmering, given the head start of overnight temperatures in the upper 80s. After a restless, fitful sleep, John Wintermann rose before his usual hour and decided on an early morning stroll for exercise before the day's full heat set in. The moment he closed his front door behind him he realized it was a lost cause, but decided on at least several blocks anyway. He lost lots of perspiration and whatever eagerness he originally had in surprisingly short order, but not before experiencing the innards of the oven that was summer's end in Southern Illinois.

The cicadas were making their noisy presence felt, drumming a faster rhythm than anything in sight wanted to march to. As he neared the branch, occasional dragonflies could be seen, but even they had downshifted to first gear, their darting slowed to something more like a glide. Shimmering heat waves rose to their own rhythms from a once vibrant street corner still advertising all these months later the charred, disreputable ruins of an abandoned house brought low by an early summer lightning strike. And from the back corner of that same lot where a garden had once provided adequate defense against such things, a tiny dust devil spiraled up and died in almost the same instant. Few flies were in evidence. It was too hot to tempt even them to venture forth, the juices long since sun-sucked from any detritus the spatsies had passed over. No rain in sight, no clouds for shade. And one lonely city truck, making its anomalous path through the town's alleys, depleting the mosquito abatement budget before questions could be raised about spraying this late in the season, not to mention in such dryness!

In his abbreviated six-block course, there was little human activity. There was, instead, ample evidence of sensible, lower-order sluggishness. A cat on its shady porch was on a lazy hunt, eyeing a robin on its own frustrated quest for prey from the front lawns of its morning flight path, its head tilting back and forth as though listening for bugs and worms rather than trying to spy them out.

The priest ambled at an ever slowing pace past a car listing on a flat tire and hugging the curb of a short, dead-end street; past a half-painted rental home; and finally past the town floral shop with its muffled start-up sounds from the green houses to its rear; till he finally came again to his rectory. The pavement in front of him was momentarily darkened by the shadow of a hawk ranging far from its territory. He was startled and peered upward into blinding sunshine, and then reconfirmed his decision to re-enter his cool rectory for some freshening up before Mass. It was so hot, he decided, that eggs wouldn't fry on the sidewalk; they'd evaporate. And the day had only just begun!

An hour after his poorly attended Mass, Annie called to ask him to mid-morning tea and was typically brief. The recent information about her prompted the illogically uneasy feeling he was about to encounter a queen bee without benefit of protective netting; but he accepted. After a quiet moment steeling himself, he drove to the mansion. The sun by now plainly in its ascendancy, walking was clearly out of the question.

Round back he knocked on the wooden frame of the summer porch screen door. The overhead fan was revolving slowly outside the open kitchen door. For a moment he thought he'd guessed wrong, Annie possibly waiting at the front of the large house and unable to hear him. But she soon appeared in the doorway and greeted him cheerily. *So far so good!*

"Hello, Father. It's *so* hot today; Please put that fan on high, if you don't mind. Just pull that little chain; I'm sure you can reach it." He could; and the airflow improved instantly. "I thought iced tea better for a day like this. We'll have it here, of course. Do you mind? Have a chair," she said, disappearing into the kitchen. It wasn't unusual, her not waiting for an answer.

He took a seat as instructed and in a trice she returned with everything on her large tray. Soon they were sipping quietly from tall glasses replete with sprigs of mint from her herb plot at the bottom of the steps. He waited for her to initiate conversation. It didn't take long.

"You wanted one of my paintings, and I've decided to give you one -- the one you spoke of. That *is* it, isn't it?" and she indicated the very painting he had admired in the upstairs hallway. It was propped against the wall in the far corner of the porch, unnoticed till then.

"Why, yes, that's it. You didn't strain yourself bringing it downstairs, I hope. It's rather large for you, I'm afraid."

"Not at all. I carried it on my lap down the back stairs."

"Might I proudly identify the artist when I display it in the rectory?" He had her unreadable signature in mind, but that went unaddressed, diplomatically or otherwise, as Annie responded promptly.

"Oh, no. I prefer not," she said, daintily taking another sip.

"If you insist," he said with regret. "But I'd *dearly* love to brag about my parishioner," the emphasis underlining his sincerity.

"No. Please remember: that would not be at all lady-like."

He could only nod silently and comply, however reluctantly. He took a slow, satisfying sip of tea, then looked up. "If I might be so indelicate: what recompense may I offer?" It was an earnest request.

"Oh, nothing, to be sure. It is my gift."

"Oh, I can't allow that," he said with determination. "You *must* accept something. I insist." His conversation with Robert Lanner loomed large in his memory. "Please," he added.

"Let's just say that it is my way of bragging about my pastor," she said with more cleverness than he thought she possessed.

"I don't wish to belittle your generosity, but I'm truly embarrassed. Such artistry should be rewarded, n'est-ce pas, Madame?"

"No, I won't hear of that either, mon Pere," she countered. "Your pleasure is my reward." She had him, and they both realized it.

And to forestall further discussion of this sort, she abruptly changed the subject. "My nephews haven't contacted you recently, have they?"

He choked at the question. Napkin to face, he coughed to clear the tea that went down the wrong pipe. Then, shaking his head ambiguously in what he hoped would be interpreted as a simple *no*, he asked in a high, strained, choking voice: "Why?" and coughed again.

Annie didn't seem to find his behavior or its timing unusual, and promptly responded: "Bless you! Well, sometime ago you spoke of putting them into my will. Wouldn't you know, that very thing came up on the phone yesterday! They seem to think we discussed it in June; but I don't recall that. Now, I admit I'm something over fifty," she said, and produced a polite, self-conscious, tiny little cough, "but I don't think I'm into my dotage just yet, thank you. I certainly do not recall any such conversation when they were here," she said emphatically. "Well, I put them straight on that, let me tell you. I just thought perhaps they had discussed the issue with you one day or another while they were here. Or since," she added, after a slight pause -- pointedly, perhaps. He couldn't be sure.

No conversation *exactly* like that! *Thank God for small favors!* "No, not at all. That unfortunate idea was my own, I dare say. Please don't think I put them up to anything or ... "

"Now, don't trouble your precious heart, Father," she said, cutting him short. "That's not what I'm thinking at all. It's just a figure of speech, to introduce the idea that they'd had the ... " she paused, in search of just the right word, " naiveté," she continued, "to pursue that fantasy. In retrospect, it seems rather humorous, actually. I fear I was too harsh, however ... no: too *abrupt*. I really must call back and clarify that."

Must've been some *conversation,* was all he could think, as he continued nursing his tea.

They sat for a while, silently sipping their cool drinks and enjoying the fan's breeze, and other than the fan motor, the gentle tinkle of ice cubes the only sound between them. Then he asked: "Are you sleeping well in this absolutely awful weather, Miss Verden?"

She corrected him almost the instant he got the words out. "Please, Father. How many times must I remind you? *Annie.*"

"Pardon me. *Annie!*" he said emphatically, and chuckled softly. "Are you sleeping well these nights? I mean, I don't know how you can. I'd simply die in this weather. I know you don't have air-conditioner one in your entire house, and the upstairs is *so* hot, I now realize."

"Oh, I do use a small fan. And I'm quite comfortable. But thank you for asking."

He was cut off at the pass. With no way to introduce the subject of the oxygen tanks now, he could only finish his drink and then mutter gratitudes: for her ability to sleep tolerably, for today's tea, and especially for the painting. Moments later, with a final request to greet her nephews on his behalf during the coming phone conversation, and having refused the proffer of another glass of tea, he bid adieu, carrying his painting to his car.

Once again he'd remove the foyer print and place another *Verden* there. And yet again he'd treat his lobby art as an anonymous wonder. This time not because he wasn't able to identify the artist, but because he wasn't allowed to. Unlike the first one, this painting did prompt inquiries. And to each inquirer he voiced gratitude to Horace Denver for such a beautiful replacement. He hoped that small fib wouldn't get back to either Horace or Annie.

August was about to slide into Labor Day and September. The city park's pool would be closing, depriving the children of their chief means of cooling off; and once again they'd be relegated to their very old and very hot two-story school where fans were a luxury and air-conditioning too expensive. Families were in the frenzied finale of last minute summer vacationing. And weekend Mass attendance was still in unfortunate, however predictable, seasonal decline. Because he had vacation time coming himself, he announced he'd be away most of the month's last week. He left after his last Sunday Mass and drove north to a classmate who pastored a Springfield parish. Besides visiting with him, he planned to tour the center of the state from that base.

It was a relaxing week of leisurely evening meals with his friend and days of sightseeing at Lincoln sites in both Springfield and New Salem, including the new *and* old Capitol Buildings. He also drove through the nearby Amish country, where one noontime he had a fine, simple lunch and bought some cheese, and where he also took time to notice the wild flowers in abundance along the highways.

The orange of early-summer tiger lilies had long since given way to an omnipresent white filigree of Queen Anne's lace. But even those, while still amply evident, were long past their exclusive control of ditch and roadside, having recently ceded the spotlight to the less tenacious cornflower. These bachelor buttons, as they're known in floral arrangements, had been interflecked for a while between the Queen Annes and occasional blue fireweeds. By now, however, even their strong blue stands were fading in favor of several varieties and shades of yellow just now beginning to vie for attention.

Ragweed, Midwestern sniffle bestower extraordinaire, had long been hiding, green and unnoticed, he knew, in many of those modest wild roadside bouquets. But just then it was some light yellow buttercups that drew attention, as well as a lot of jaunty yellow goldenrod, which like ragweed, is also a noxious pest.

Passing field after field, he eyed the miniature stone obelisks demarking each patchwork's boundary. Where those were missing or crumbled, iron rods stood in replacement at the corners of the fields, often sporting empty quart oil cans or anything similarly possessed of a narrow spout and some depth so they'd stay securely on the rods and clearly visible to tractor drivers, to stop them from plowing or planting someone else's field *or* encountering a ditch.

He also saw Black Eyed Susans and helianthus, Jerusalem artichokes and stick-tights. These latter were part of every colorful display fronting the fields, and they rendered up-close floral identification difficult. A hike along the road shoulder was the only sure-fire method for that precision, but the nuisance on socks and trousers of beggar lice, or stick-tights -- Nature's very own Velcro -- causes second thoughts about such excursions.

Also, only recently had he begun to see *entire fields* of yellow dotting Illinois farms, and even more recently still had he learned these were cash crops of canola, not wild stands of any very bad news weeds. The gorgeous bejeweled fields and roadsides were much more in evidence on rural highways *and* far easier to savor at the slower speeds mandated there -- more reasons still to avoid Interstates. Such sights were immensely enjoyable.

He was also much delighted with Allerton House, the large estate outside Monticello whose spacious grounds were filled with almost endless, meticulously manicured flowerbeds of experimental varieties. There was also a surprising number of outdoor sculptures, the most interesting of which being *The Sun Singer.* He continued sidestepping Interstates, stopping whenever scenery or people intrigued him. The region's rich soil and crops never ceased to enthrall him. Time and again he pulled over beside particularly lush fields to sit and stare at the immense loveliness of it all, and often as not he found himself at prayer beside these Midwestern wonders of the Lord.

He returned Saturday morning, and with only a few days of summer freedom remaining, he put finishing touches to a blessedly brief reflection for his weekend Masses concerning the *rhythms of responsibility.* He opened his pile of mail, and settled back into his own personal set of responsibilities: the care of cat-licker souls at Saint Helena's.

He awoke Sunday to the threat of area-wide thunderstorms but managed to finish both Masses without any rain. Annie had sent word via neighbors to bring the Eucharist the next day, so he made a mental note to include several other shut-ins on Communion calls after Labor Day morning Mass. Ordinarily that was a monthly First Friday routine, but the holiday was a good time, and close enough

145

too, he hoped, to September's first Friday, to avoid offending any older folks still adhering to that pietistic devotion to the Sacred Heart.

A storm finally hit just after noon, disrupting many a family outing. Later there were reports of a small tornado in the Missouri Boot Heel at that same mid-day hour. It failed to follow the usual pattern and slip over into Illinois, however, staying in the Cape Girardeau area where it caused no loss of life. Once again, though, there was no relief. It was still boiling, with even higher humidity. So he spent the rest of that sticky September Sunday reading inside, air-conditioning cranked up to maximum.

The next morning he slipped away for battercakes du jour at the truck stop. Elsewhere these breakfast delights are called *pancakes, hotcakes* or *flapjacks* -- sometimes even *flannel cakes*. But the truck stop insisted on *battercakes*. Not even *griddlecakes* was distinctive enough. *Battercakes* it absolutely had to be. By whatever name, for over two months the heat had precluded his indulging. But that morning, upon awakening early, some deep-felt need had activated his salivary glands, and with the holiday morning Mass an hour later than usual, he decided he'd have time for that special decadence, especially when early temperatures wouldn't yet overtrump his palate.

So he went out to enjoy three of the large culinary creations with oceans of melting butter and warm syrup oozing almost over the edges of the plate. Cold milk – always *cold* milk with *hot* battercakes -- completed the delightful breakfast. And conversation with the other early-rising regulars was companionable.

He learned nothing new about Algoma, but there was thunderstorm information from the experts at the counter, the place where anybody who is anybody among truck stop habitués sits to partake of gourmet fare. Surprisingly little crop damage in either Illinois *or* Missouri had resulted. The good news was that farmers got needed moisture, *to keep from gettin bone dry,* according to one of the amateur agrono-meteorologists. The bad news was that no front had come through, so the sticky, miserable weather was hanging on.

As he was leaving he got a friendly reminder to *get a good seat for the big parade,* which this year was to include a record number of bands from area schools. He'd already decided that: Annie's front porch. Hers was therefore the last Communion call that Labor Day.

A surprising amount of work goes into such parades. Small towns still cling to a long-standing tradition of parades for holidays and moments of civic importance. Belleville at one time boasted more parades than -- anywhere, he supposed. He couldn't remember for sure; but they used some such superlative language.

Algoma had enough of them; that he also knew. And local preparations were extensive: another reason to absent himself the week previous, to avoid being co-opted! He preferred watching the finished product, if that. Some parades he avoided altogether. But this Labor Day would be spent, God willing, on Annie Verden's porch. Since all parades pass her front door, her porch is *a prime and painless place to parade peruse,* he had once told her, though surely the essence of that wasn't news to the duchess, however much the phrasing might be.

Annie's porch offered several advantages. There'd be a cool drink; and by parking on the side street he might cleverly come late and leave early. He couldn't openly complain about a parade's length, let alone that it was boring or that flies were eating him up. But some other excuse -- parish business -- *no, too obvious and over-used; he'd need something else* -- might allow an early exit after demonstrating appropriate and proportionate civic loyalty via his symbolic appearance in the highly visible venue of the duchess' throne room. All this, presuming, of course, that Annie wouldn't insist it was his fullest duty to stay the course. Which is what she'd almost certainly do, and which would therefore in the end determine his schedule.

So, the most he could bank on was arriving just beforehand. Which is what he did, coming barely minutes before the first entry hove into sight.

He had, in fact, barely enough time to give Communion and settle into a large wicker chair in front of the throne swing, iced tea and hand fan in hand, when a high school band appeared, brass blaring and in tune to boot, at just moments after two. As it turned out, the experience was rather pleasant. Tea and conversation flowed well, thanks to a chatty duchess and the icy pitcher on the low table to his right so he could serve them both. He stayed to the bitter end, till the last horn tooted and the last tractor belched by. It was mostly small talk, but in the end Annie pledged him an apple pie, made with *real Murphysboro apples fresh from the Apple Festival.* He should come for it in a few days. He promised to.

There had been no opportunity to bring up the oxygen, though he tried a few sallies into that territory. No nibbles; no luck. But, rehashed in memory after the last salutes from the fairgrounds' fireworks had rumbled to silence that night, it had been a pleasant Algoma afternoon. Nothing more, but, thankfully, nothing less, either. He slept contentedly, relieved, as he always was when an encounter with Annie had gone well. Or at least not poorly.

CHAPTER XVII

The next morning there was a mist easing off the rectory lawn courtesy of a pre-dawn sprinkling.. *Thank God for automatic timers, or my lawn would be a dead, brown eyesore.* Father John was glancing out his bedroom window. He began groggily putting himself together. A mourning dove cooed dolefully nearby. A sound so common, it's often just background noise, but this morning it got his complete attention, the mournful call momentarily sad and foreboding, like a lonesome train whistle in the prairie night. But it slipped from awareness and he was soon retrieving the newspaper for perusal over pre-Mass coffee.

Nothing startling. The usual ambivalent outcomes in both national politics and the Saint Louis sports scene. Riverboat attendance was up slightly: good news for embattled East Saint Louis, receiving, as it does, tax revenues from the largest of the Mississippi's floating casinos. Late crop returns showed a large but not bumper harvest. And the entire Midwest posted record temperatures during August. *So, what's new?*

After Mass he braved the cauldron they'd been boiling in nearly forever and drove into the countryside to discuss -- at their invitation -- writing the parish into an older couple's will. On his way through town he was taken once again by the vast assortment of *bumper art,* as Herb had once styled vanity plates and bumper stickers in the Smile. He found their popularity hard to understand. Long since beyond political advertising and random touting of special issues like ecology, now it was just cute stuff, and only a bit of it funny. His favorite was a yellow smiley face with a single eye in the center of its forehead and bearing the motto *Mutants for Nuclear Power.* Just before he eased out of town a smallish bumper sticker only tailgaters could read caught his eye: *Made in Indiana by the Indians.* It didn't rate a ten, but few did.

Owners of vanity plates, he'd learned from a friendly politico in a nearby state, were routinely lumped into a pool from which a few were chosen for state income tax audit, on the assumption that if there were discretionary funds for such frivolities, the state wanted to discover just how big those funds were. Their inflated cost had already kept him from indulging; this new information guaranteed it. If some smaller state had figured that out, Illinois darn well had!

But there were still occasional grins. *GN HUNTN* had been seen on a much-used out-of-state pickup sporting not one, but two, gun racks mounted over the rear window. *PILZ ETC* adorned Fred and Frieda's Chevy. And there were others he routinely saw: *WILD MAN; RUN 4 IT; 2N2R4; R U SURE; DOPEY.* His favorite was *H2O NO,* on a young fellow's badly banged up clunker. It had been hit quite late one Friday night in some previous summer by someone who received a DWI for it.

On the five miles of dusty road to the Keller farm, he noticed numerous other things. From atop a small hill he could see trees everywhere, three hundred sixty degrees around. Trips through the West's vast treeless stretches had convinced him of the peaceful blessings of his own wooded Illinois horizon. Nearer, there were tasseled cornfields, gleaming a solid, misty gold gone just beyond its peak of color, he realized. Fall had already begun to dry the stalks and dull the fields back from their zenith. Quiet, green stands of soybeans were dotted throughout, he knew; but the corn's height gave the illusion of an endless golden river. His lowered window brought in, along with a blast of heat, that yellow river's sweet, overwhelming aroma: the sure smell of success, the odor of harvest and attar of assurance that the year's bills would be paid with something more to set aside against the inevitable lean year every farmer knew loomed just over the fiscal horizon. The only imperfection: a thin trail of black smoke from behind a barn in the distance. Though not born to the land, such sights and smells never ceased to speak to him of home.

The gentle old couple was gracious and wanted to help the parish they'd always felt to be their spiritual support. They'd be as generous in death as in life, having quietly helped many local endeavors even while rearing seven children, several of whom they'd sent to college. Father John was pleased to spend a few delightful hours with them, essential matters dispatched within minutes once they'd put their collective minds to it. Then they reminisced about family, farm and parish, all interwoven in an intricate and inextricable pattern, each thing bound to every other. Though life seemed more complex now for younger folks, the priest was happy to know that for them it was simple. He was also happy their generation was his own.

While returning, his car struck a small bird. Something strange had been happening with the birds all throughout the heat wave. This was the fourth bird his bumper had dispatched within just the last month, all of them flying low across his vehicle's trajectory. Three had died on open highways; this was the only one constrained into low-level flight by corn phalanxes flanking the road. *Is this connected with ozone depletion?* He recognized the silliness of that almost instantly, and wondered where the thought had come from? Then he frowned to realize that most other natives, faced with similar epiphanies, would probably not be at all uncomfortable doing in a bird or two.

Still, it was eerie; ominous. Birds are quite agile. He understood the occasional dove splatting onto a truck windshield. But for any other bird -- let alone four in so short a time -- flying so unusually low and victims of ordinarily avoidable machines! It seemed unnatural. *Some weird harbinger?*

He was barely inside his rectory, unopened mail in hand, when the phone rang. *Could he come to Saint Luke's?* "Coonie's had a stroke, Father," was the desperate plea from Julia, Conrad Eversman's wife of many years.

"Certainly, Jules. Be there within ten minutes. Stay calm. I'm on my way." He hung up, dropped the mail onto the end table, and headed back out to his car even as he realized he hadn't asked about his friend's exact condition.

Conrad and Julia were the sort of rock-solid older folks most priests are delighted with as parishioners. Coonie had sold him several cars from the town's Ford dealership, which was now run by his two sons. Jules had taught kindergarten in Algoma forever -- easily several years longer than she'd been married to Coonie -- so, forty-plus. They had both formally retired a few years earlier and were often traveling someplace or other within the country -- *always the U S of A, never anywhere foreign,* Coonie would say. He'd explain at great length, if you let him, that it had to do with the economy and *buying American.* The trips were seldom more than two weeks in length and almost never exceeded three a year -- four only once in an exceptional display of flexibility.

When they weren't away, they could be seen at virtually every community volunteer effort, most charitable events, and nearly *all* Saint Helena's services. She was sacristan and he head *greeter,* parlance hardly anyone was used to, except Sister Joanne, Saint Edward's religious educator, who introduced the term to her pastor and Father John, and who kept correcting them when either might slip from force of long-established habit and say -- God forbid -- *usher!*

The couple had been so careful to eat properly and take lengthy walks in the park most days at the coolest times. *Don't want to walk or run along roadways, Father,* he could hear Coonie say. *Too many exhaust emissions -- I should know, right? Walkin or runnin like that, you're starved for air. Worst thing you can do is clog up your lungs with that bad air. Get cancer fer sure.* What irony to be hospitalized now despite such precautions!

He loved them dearly, these two precious cat-lickers, and on entering the E. R. it almost broke his heart to see the fear etched onto Julia's face. While they worked on her husband she was confined to the waiting room across the hall, thoroughly distraught. He calmed her as best he could, then crossed the corridor toward four or five hospital personnel hovering round their patient. A nurse motioned for him to wait momentarily, then whispered: *He's not in extremis, Father.* So he said something deliberately lighthearted to Coonie about being there for him and Jules and that he'd see him upstairs shortly. Then he returned to the waiting room with the reassuring news.

He anointed him some thirty minutes later in the presence of his wife and both sons, the boys having by that time arrived from the dealership. He stayed on another twenty minutes to ensure things were sufficiently in hand. Thank goodness Coonie was suffering no speech- or motor-impairment. But after a second stroke several hours later, the priest found himself by early evening back in the second floor room of his now comatose friend.

In fact, as evening turned to night, he remained at Coonie's bedside, having sent everyone else home for as good a sleep as possible, under the circumstances. "I'll do my catching up after Mass tomorrow. Got a light day." He fibbed as convincingly as he could.

He spent the long night holding his friend's hand, the only sounds the hiss of oxygen and his own soft whisperings. Between occasional dozes, he had the whole night to think about the Eversmans and his other Saint Helena's cat-lickers. Heat lightning danced the distant horizon of the moonlit, cloudless sky, telling him the weather hadn't eased its onslaught one bit and the dry spell wasn't over. The fall leaves might be robbed of their spectacular colors, much like Coonie, robbed now in his own autumn.

As it turned out, there was no catching up on sleep for Father John after Wednesday morning Mass. He had just walked into the rectory when Feldspar's called to tell him he was needed again at St. Luke's. He thought immediately of Coonie and asked if it were he.

"No, it's Annie Verden."

"Is she all right?"

"She was found dead in bed this morning, I'm sorry to tell you, Father."

"I'll be right over," he said, and hung up to hurry back to Burger, planning to get more details once he arrived. Odd though, because she was not only alive on Monday, of course, when he left after the parade, but apparently in good health as well.

As he drove over to Burger he prayed silently for Annie, and included Coonie for good measure. *I'll check on him after I anoint Annie.* He removed the oils from his glove compartment as he drove and laid them beside him, and in one motion picked them up as he opened the car door at the Emergency Entrance.

From force of habit he rushed into the hospital, finding himself questioning his haste once he came alongside Annie's body. They had left him alone with her in one of the several emergency rooms, so he took time to settle into a prayerful composure. He looked around to determine that he was indeed alone, and decided to do the longer form of anointing as a gesture of respect to a long-time parishioner. It was not until he actually touched her forehead with the oil that he realized she'd been dead a good while, her skin quite cold, but surprisingly pink. *Too much sun at the parade?*

After a brief moment of prayer following the sacrament, he stepped outside, chatted briefly with the nurse and went in search of Coonie. He'd seek out Mr. Feldspar, Maisie, and perhaps a neighbor or two of Annie's as well -- but later. First, Coonie! He was surprised to be met by the entire family as he entered the room, and immediately feared the worst. But they were smiling, and so was Coonie, though weakly, as he lay silently in the hospital bed.

"He awoke shortly after you must have left, Father. They called us and we came right over. We were going to go to Mass this morning, but this seemed more important. His speech is okay and he can move everything. But he's very weak, and we're not letting him talk much."

"Thank God!" the priest said to no one in particular, and then turning to the patient, he patted his hand and said: "I won't make you talk either, Coonie. But I *will* be back tomorrow when I hope you *can* talk some. Meantime, just rest." He gave him a slow, deliberate blessing and included everyone in the room. Smiling broadly at Julia, he stepped out into the corridor and made immediately for the emergency room exit, without telling them what had really brought him back just then.

On his return he mapped out strategy, and made straight for the funeral home as he entered town. Junior was there, and they were quickly conversing about the circumstances of Annie's demise.

"In answer to your question, we found her lying in bed rather peacefully. Maisie Brown called us. She was the one who found her."

"Had she been using her oxygen tank?"

"Can't say for sure, Father. Wasn't hooked up when we arrived, but Maisie could have disconnected her. You should ask *her.*"

"Well, I intend to talk to Maisie. Anything out of the ordinary at the scene?"

"Nothing that struck me, why?"

Just curious. She didn't seem to be ailing, is all. What do you make the cause of death to be?"

"Old age, I suppose. Why? Think we need an autopsy? Don't believe the coroner's planning one, the way she talked at the hospital."

"Don't know. Just seems surprising. Anyway, thanks for your help, Larry. Tell your father hello; I'm sure I'll see him at the funeral." Then, just before he got to the door, he turned again to the funeral director: "Just in case we *might* need one, an autopsy, that is, could you wait with the embalming? I mean, there's no rush, is there?"

"Not really. Suspect something, Father? Foul play, or ... anything?"

"Can't say I'm certain about anything, Larry -- it's just a weird feeling. Give me a few hours. I'll be in touch by mid-afternoon. Ok?"

"Fine by me, Father. Meantime, I can contact her relatives. Nephews, I believe. Somewhere out east."

"That would be Pittsburgh. Yes, nephews. If you need a phone number, I can probably come up with one; just call the rectory."

"Fine. But no later than mid-afternoon, ok?"

"Deal!" he said, and walked out. On the outside chance Maisie was still at the mansion, he went there and, sure enough, found her cleaning the place. It wouldn't occur to the girl not to complete the day's work *or* that there'd be no one to pay her, what with Annie gone.

"Maisie, you all right?" the priest asked when she opened the front door. "Finding someone dead is usually a big shock."

"I be all right rev'ren. Jes sad about Miss Annie."

"You found her when you arrived this morning, is that right? About when was that, Maisie?"

"Bout eight o'clock or so. I called Mr. Feldspar right off when I could tell she done passed." *So, she* can *use a phone!*

"Tell me, Maisie, did anything look peculiar; wrong; out of place?"

"No, sir, sure didn't. The fan was on, blowin a little air over her."

"Were the windows open or closed, or the bedroom door locked? Or anything at all strange?"

"No, sir, nothin strange. Don't think she ever locked anything upstairs; and there was one window open ... no, two. *Both* windows in her room was open; it was actial kinda nice when I come in."

"Did she use her oxygen overnight?"

"Why, yes she did, rev'ren. I took that off her and turned it off, cuz that kinda stuff be dangerous."

"Mr. Feldspar said the mask wasn't on her. That's why: *you* took it off! Did you do anything else? When people die, police or coroners sometimes ask about things like that -- in case there's any question about the death. If they ask, please tell them everything you're telling me. So, did you do anything else besides take the oxygen mask off and shut it down?"

"No, sir, I didn't." She was emphatic. But after a thoughtful pause, she added: "But, you know, sump'n peculiah: her color! Wasn't like her, you know. She protected herself from sun and stuff."

"Color?" He remembered the pinkish cast he saw at St. Luke's.

"Yeah, kinda light red ... like sunburn. But that don't seem right. Was she out'n the sun, rev'ren?"

"She and I sat on her porch for the parade, Maisie. I remember seeing that same thing at the hospital this morning, and I thought it must have been the sun too. But now I think back, we were well shaded Monday. And, like you, I doubt Annie would have carelessly gotten much sun *any* time, really. That *is* puzzling."

"Yes, sir, it be a puzzle all right! You know, I sure gonna miss cleanin here. She had some right nice things downstairs here, and all them pretty pitchers up the next floor."

The priest agreed, but his mind was racing ahead. He said good-bye and offered to pay her himself. Maisie thanked him; she'd contact him if need be. She'd also reconsider the rectory job now.

Father John headed back to Saint Helena's in confusion. About the only sure thing now was that too many things didn't fit. And, what with a dead body, he might just have finally found himself nose deep in a humdinger of a mystery, right here smack in the middle of Algoma.

CHAPTER XVIII

There were messages. Only Lafe Skinner's needed attention. It didn't spell out why he'd called. Father John needed time to think, and he hoped a quick conversation was all that stood between him and some time to himself.

But it wasn't to be. Lafe just had to see him, and Father John didn't feel like explaining about his deep-felt need to tackle this tempest, teapot variety or otherwise. So he went out to Lafe's place, hoping to keep the visit short.

Just moments inside the Skinner home he was invited back outside to the rear of the house ... to see Lafe's *flowers!* Father John felt snookered. Only then did he remember a conversation from the beginning of summer concerning the orchids Lafe was cultivating, and how perhaps some of them might grace the church at special times. While the priest didn't feel he could refuse to see them now, he was nevertheless already thinking how to abbreviate the moment.

As they entered the small greenhouse just beyond the kitchen door, the priest gasped first at the blast of moist heat and then, as he got accustomed to the intense artificial climate, at the beauty around him. "After your air-conditioning, this takes getting used to. But what beautiful orchids!" he said.

"Actually, they're not all orchids, Father. There are lots of other exotics here too, like this pigtail, for instance."

"Pigtail? Is that really the name you'd use in a flower show?"

"No, course not. Technically it's an *Anthurium Scherzeranum,* but it'd be labeled a *flamingo flower.* It's native to Central America."

"What's that deep red part?"

"Its spathe. Calla lilies and jack-in-the-pulpits have one too. This one's *reflex* since its edges turn in sort of protectively. And, to be precise, that color is *scarlet.* You have to shade it down to get *red.* Even more for *pink.* But pigtails don't have those shades. Just scarlet. The real *flower* on this plant is within the spathe on that spikey thing, the spadix. Altogether beautiful, don't you agree?"

"Certainly is. How can you grow them here, these tropical flowers? Is it just a question of the right temperature?"

"That's part of it, of course. There's also the humidity. Notice how close it feels? Then too, there's the methane -- my secret weapon, so to speak. Take a deep breath. Can you pick it out?"

"Well, now that you mention it, there *is* a distinctive smell."

"If you were an old miner like me, Father, you'd catch that immediately. And you'd call it firedamp. Leastways, that's what we called it down the mines. Comes off the coal. Here I need an elaborate system to get it from my cows. I'm trying to find a way to get carbon dioxide in here too. You know, plants feed off that and give back oxygen in return. In nature they seem to get enough from all us animals. And, of course, in turn, plants go a long way toward supplying us with what we need to breathe. A good trade! But for a greenhouse, the very need to keep it enclosed makes carbon dioxide scarce. So far, I haven't figured a good way -- an economical one -- to get lots of CO in here."

"Pardon, Lafe. You don't want CO. You want CO_2. *That's* the formula for carbon dioxide. CO is carbon *monoxide* -- like car exhaust, and toxic to us, of course. To plants, too, *I think*. It's part of the pollution damaging big city greenery, I believe. But I'm less sure about plants' intolerance to it. Anyway, unlike methane, you can't identify it by smell, I'm told. Except in car exhaust."

"Silly mistake. I know better," he laughed. "You must know your chemistry, Father." The priest's eyes glazed over as he suddenly got lost in thought. Noticing, Lafe asked: "You all right, Father?"

"Yes. Pardon me. When you mentioned chemistry, something clicked in my head. I've got to check something -- important! And right away! Please pardon my abruptness. I promise to come back. I really do want to know more about your flowers. But I've simply got to go now. On top of this, Annie Verden was found dead this morning. Hope you'll understand." He turned and walked outside toward the driveway, leaving Lafe in a metaphorical cloud of dust.

On the way he decided to call Ed Wallen at the Newman Center. Fortunately he was in. "Ed, remember those two scientists? ... Yeah, those two! Can you tell me the faculty members they talked to? I need to get hold of them as soon as possible. I'll explain later. Just give me phone numbers ... You can? Great." In another minute he was dialing the office of a chemistry professor.

"Doctor Day, I'm Father John Wintermann, a friend of Father Wallen's. I'm pastor in Algoma and hope you can help me." He was speaking rapidly. "In late June you talked to two professors from Pittsburgh about the properties of some gasses. You recall? Good! What *exactly* did they want, if I may ask?"

"Why?" the disembodied voice on the other end asked.

"Well, it's long and complicated. Their aunt has died suddenly. It's not exactly clear, but it may not have been from natural causes. I'm trying to narrow down the possibilities with a couple of quick questions. I think you can help."

"I'll try. As I remember it was all rather theoretical, about if and how one could relatively easily store certain gasses under pressure."

"Would one of them have been carbon monoxide?"

"Why, yes."

"Well, what'd you say? Can it be done?"

"Well, yes -- and easily enough, in fact. We discussed how to go about it. You sure you want the technical explanation?"

"That won't be necessary. What you've already told me is helpful. But can you also tell me, did you perhaps even give them someone to talk to in St. Louis? Maybe at Washington University?"

"Why, yes, as a matter of fact. Doctor Harry Fitzgibbons at Wash U."

"You don't happen to have his number?"

"I do," he said, beginning to fumble with his Rolodex. He soon supplied it, and Father Wintermann cut the conversation short, pleading truthfully enough that time was of the essence. He thanked him, hung up, and immediately dialed St. Louis. Once again he found someone at work.

"Doctor Fitzgibbons, you don't know me. I received your number from your Carbondale colleague ... Yes, Doctor Day. He said you were classmates. My name is Father John Wintermann, pastor at Saint Helena's in Algoma, Illinois. I hope you can help with a most important matter. Some time back you spoke with two professors recommended by your friend. They were from Pittsburgh, Pennsyl... Oh, good. Can you remember that discussion. As I'm told, they wanted to pressurize some gases -- in particular, carbon monoxide."

He remembered.

"Did you tell them how to successfully do it?"

"Yes. Actually it's not that hard, though I said I couldn't imagine why they'd want to. They told me it had to do with some experiments at the eye institute of their university."

"And you believed them?"

"I had no reason not to. Why?"

"Yes, of course. Did you in fact put some in a container for them?"

"No. I can't do that very easily here."

"So, they were unable to do it?"

"Well, I told them of a colleague here, but I thought -- and still think -- he was on vacation at the time."

"You *think?*"

"Well, I never checked it out."

"Could you? Or could you give me his name and number?"

"Yes, I suppose so. Let me talk to him and call you back."

165

"Thank you, Doctor. But before you do, do you happen to know the effects of carbon monoxide on a human body? I mean, besides it's being toxic? Like: what specifically does it do, besides affect breathing and oxygen intake?"

"Other than that it's poisonous, I don't. I'm a chemist, not a pathologist, Father. Why? What's all this leading to?"

"Well, it just may be that someone got exposed unintentionally to some of that. No one's sure just yet. We're just beginning to check. But thanks for offering to contact your colleague. I'll await your call."

Within ten minutes he called back. The man *was* vacationing then and never talked to the brothers. The priest thanked him, hung up, and sighed.

It seemed an impasse. There was probably a suspicious canister. *Did it hold carbon monoxide; and what was its origin?* There was no definitive way at the moment to be sure, let alone tie it to the brothers. Given their small sheriff's department, the priest assumed that checking all possible suppliers would take much too long. They probably couldn't detain the brothers *and* would thus lose any element of surprise.

Father Wintermann then called the Pathology Department of Washington University's Medical School. The chemistry professor had supplied that name and number. In another minute he was explaining himself all over again.

"Thanks for humoring me, doctor. Please tell me what a medical examiner would typically find upon examining the body of someone asphyxiated by carbon monoxide. Something in the back of my mind says that it affects skin color. But I can't be sure."

He listened intently for about three minutes, then thanked the man. "The county authorities here may well be calling you to verify the details of our conversation. You see, we have a body here with those symptoms, and so far no one has thought to check for anything, let alone CO poisoning. I'm going to call our coroner now and suggest she phone you. She's not a pathologist or even a physician, and may well want to talk at length. Will you be by your phone?" Assured that he would be, Father John thanked him and hung up.

His call to the coroner set that part of the machinery in motion. She agreed to contact the Saint Louis doctor immediately. And Father Wintermann told her that to save time he'd alert Feldspar's to keep the embalming on hold. He'd also tell the sheriff what was developing.

Both calls were completed in five minutes. But at the insistence of the sheriff, the three of them, priest, sheriff and mortician, decided to meet as soon as they could at Feldspar's, where, after all, the body was.

Annie was still showing the peculiar coloration Father John noticed earlier, though it seemed slightly diminished. While awaiting word from the coroner, the three conversed more about her nephews than the deceased.

The brothers were catching the next plane to Lambert Field and would rent a car. They hoped to arrive in Algoma that evening and told Mr. Feldspar they'd come directly to his place. That is, unless they got in too late. In that case they'd stay near the airport and come out the next morning. Either way, they'd talk to Mr. Feldspar that evening.

"No," Larry told the priest in answer to his question, "I can't tell you much about how he reacted. I don't recall anything in his voice except concern. But, truthfully, I wasn't listening for anything else."

The three had been together perhaps fifteen minutes when the coroner called. She had scheduled a post mortem. The local pathologist would be coming soon, and she herself would be there momentarily.

In the meantime, the sheriff said, he'd go to the mansion to confiscate *all* oxygen tanks, "And there'd better be three, let me tell you," he said.

At that moment the priest remembered the Washington University chemist whom he had yet to phone. "While I'm at it, sheriff, do you want me to ask how we go about determining if the tank Maisie unhooked ever contained carbon monoxide?"

"Do it," the sheriff said. And, pausing in the doorway, he said to the funeral director: "If you talk to those two *before* we get autopsy results, stall them with funeral arrangements. By no means allow them near the body *or* mention the post mortem. Not till we know more."

"Understood," Larry said, as Father John reached for the phone. He asked Larry to listen in so they could both take notes on what to do with the tank or tanks. In ten minutes they had what they needed.

"I remember learning somewhere what the symptoms of CO toxicity were, but I couldn't be sure. Thank goodness I was essentially correct about skin color. But I didn't think to ask just now how long that lasts. Think we should take some color photos?" the priest asked the funeral director.

"Can't hurt," the mortician said. He had a 35-millimeter camera in his car and went to get it. Within minutes they had the pictures. The symptoms were still visible on Annie, and the camera would capture that. They also noted the time of the photos. Mr. Feldspar promised to get them developed the next day.

When the sheriff returned, at Father John's instigation they discussed how to handle the large amount of so far circumstantial evidence. They'd have to tie the tank to the nephews or they wouldn't have an airtight case. Father John passed over the pun without comment. But he noted their apparent consensus as to the nephews' highly probable guilt.

On the spur of the moment, Father John thought of St. Louis University. "It's an outside chance, but maybe they went across town when they couldn't get what they wanted at Wash U." It was another ten long minutes and several transfers before they got through to someone in the SLU Chem Department. It was someone who not only remembered talking to the nephews in June, but he also had been able to satisfy their needs. They'd gotten very lucky!

The man also said he hadn't supplied any oxygen -- that requires a physician's prescription. *They probably got that in Illinois.* But he *had* provided the canister of CO.

Perhaps the professor could remember a serial number or some distinctive mark, the sheriff suggested. Father John asked, and the man said he'd check his records and call back *today or tomorrow.* But he *had* clearly labeled the old oxygen canister as containing CO.

"How, exactly?" the priest inquired.

"With the handwritten caution *CO* on a large gummed label."

"Which could be removed?" the priest asked.

"Well, yes. But why would they want to?"

"Just speculating," the priest replied. "Are you certain it was originally an *oxygen* canister?" *He was.* After some pleasantries Father John hung up, with mutual assurances that return calls were welcome.

Time to turn the heat up. They agreed the trap had to be set carefully. They'd need a lawyer to ensure evidence was processed properly. Pat Kelly was out of the question. As Annie's personal lawyer, he might muddy legal waters. The State's Attorney was also not a good choice. This hadn't yet become a criminal proceeding, but soon might.

They settled on retired Judge Hugh Monroe, and the sheriff put in the call. They all breathed a sigh of relief when he agreed to come aboard.

CHAPTER XIX

As it turned out, the nephews couldn't arrive till the next day, which gave the wheels of justice time to grind out both an autopsy and an examination of the tank that had, by Maisie's testimony, been used the last night of Annie's life. It *had* contained carbon monoxide, *not* oxygen. In fact, they found some CO still inside. And Annie had indeed died from inhaling a lethal quantity of it.

But finding the SLU Chemistry professor's label might prove to be another matter! And there was at least one other loose end: the professor identified the tank's manufacturer, but he couldn't supply any serial numbers or distinctive markings –- it was just like all other such tanks, he said, except it was obviously older and nicked up a bit.

"Maybe the canister was supposed to be empty by the time the body was found, with no way to check its former contents -- if we'd even think to," the priest mused aloud, mostly for his own benefit. The sheriff nodded in silent agreement. The two were once again huddled, this time in the sheriff's office around 9:10 Thursday morning. The brothers had yet to turn up.

Maisie had already explained to the sheriff why something might still be in the tank. As she'd told Father John earlier that summer, Annie had the rate of flow slowed down to make the gas last longer. Father John had mentioned hearing that once or twice from Maisie, and she had reconfirmed it just moments earlier.

"It's all falling into place," the priest volunteered. "Just so the fish doesn't spit out our hook! We still need the doctor who gave the oxygen prescription. And a comparison of the tanks might show some discrepancy between the genuine tanks and the St. Louis one." For, as it had turned out, all three tanks were from the company that had the regional monopoly. The sheriff thought he'd have a comparison soon.

They were talking about how to finalize things as Judge Monroe walked in, true to his word. A quick conversation determined that everything was fine so far. But there was more to do.

The sheriff went to check the canisters, which by now were in his office and, as luck would have it, *none* of them were in mint condition. But sure enough, the one with the CO had a different code sequence and looked older and more roughed up than the other two. That was promising. But, how to find the prescription writer -- without, that is, asking dozens of local MDs? Maybe the nephews would provide that, wittingly or otherwise.

While they were puzzling over that, the phone rang. Larry Feldspar said the nephews had just arrived and he was stalling. The sheriff said they'd be on their way.

Larry was discussing funeral arrangements when they arrived. The priest's appearance brought a grateful look of recognition to the brothers' faces; but when the sheriff and judge were introduced, their look changed to surprise and puzzlement. Father John gently broke the news to them that their aunt had died under suspicious circumstances, and their demeanor immediately changed to apparently authentic anger, shock and outrage.

"Was it foul play?" Joe asked.

"Well, we aren't absolutely certain yet, but that's probable," the sheriff said. "Your aunt died from breathing carbon monoxide from one of three supposed oxygen tanks. I'd like to know where you got them, because I understand they came into the mansion by your doing." He sounded ominous. Father Wintermann was watching the brothers closely, and so far they seemed to be playing their parts well.

"We got the prescription locally," Anthony explained. And then, as his face changed visibly, he added: "Are you implying we brought this about?" He was now clearly outraged and defensive.

"We're simply trying to determine how this happened. And you're logically among the people to question," the sheriff said coolly. "Who wrote the prescription?"

Struggling to contain himself, Joe looked at his brother and said: "Wasn't it Doctor Perkins, Anthony?"

Yes was the terse reply. *Here on the square.* As though the locals wouldn't know!

"Easy enough to check! Father, would you mind calling the doctor while I continue talking to these gentlemen?" the sheriff said, coolly.

Father John left the room and soon was able to verify that it was indeed Doctor Perkins who had not only written the prescription but had also suggested that three tanks, small ones, be gotten initially. And he was prepared to extend the order, if needed.

173

By the time the priest returned with his report, the intensity level had greatly increased. The judge had been listening quietly so far, and the sheriff had just been told that the brothers were present when the first two tanks were delivered to the house right from the pharmacy. The third tank was unavailable and would be delivered later.

Convenient, the priest thought. *And a nice touch!* But he couldn't help wondering why Fred and Frieda hadn't said anything about that. That was the sort of news they loved to circulate, at least to people on the inside, like himself. He broke into the conversation.

"Pardon me, sheriff," and turning to the two, he continued, "but may I ask: did you get the oxygen at the drugstore here in town?"

"No," Joe answered. "I thought it peculiar, but we were directed to Burger. To the Rexall drugstore."

"Why there?" the priest blurted out, without thinking.

"That's easy, Father," the judge cut in, speaking for the first time. "Dr. Perkins is a Methodist and often by-passes Fred and his wife because they're Lutherans. It's a sticky wicket no one talks about, but a bunch of us figured it out quite some time back." He smiled a rueful little smile.

"Well, I thought it strange," the priest continued. "*And* I don't think Doctor Perkins is Annie's physician..."

"Sorry. We didn't know that," said Joe, looking at his brother who was nodding in agreement. "And *as I said,* I thought it odd to have to go to Burger."

"So, when was the third tank delivered?" the sheriff asked.

"After we left. Our aunt confirmed its arrival on the phone a few days after our return to Pittsburgh."

Probably difficult to check -- convenient again, thought Father John, and he looked at the sheriff, hoping he'd bring up the CO tank from St. Louis. When the sheriff shot back a quizzical look, he said: "Did you perhaps want to talk about St. Louis, sheriff?"

"Yes. Thank you." Turning back to the young men, he asked: "What did you do with the CO tank you obtained in St. Louis?"

Both brothers bristled, but Anthony spoke, pointedly: "It went back with us to Pittsburgh. It was for a doctor at our university's eye institute there. Why? And how did you know about that, anyway?"

Joe added, deliberately, tersely, and with barely disguised outrage: "Are you suggesting we murdered our own aunt?"

"No," said the sheriff calmly and with equal deliberateness, "but your story will have to be corroborated or you *will* become suspects."

"Well, let's just do that right now," said Joe, all traces of civility gone now from his voice.

The judge agreed to call the name and number in Pittsburgh that Anthony wrote down for him, and left to make the call. The icy quiet in the room while he was gone, because there was no conversation the whole time, made the difficult moment almost impossible.

When he returned, the judge acknowledged that the professor at the eye institute had indeed confirmed the brothers' story, down to a description of the tank, the hand-written label and the

175

manufacturer's name on the obviously second-hand tank he was using for his research at the institute.

So -- where do we go from here? was all Father John could think, and the look on his face conveyed that perfectly to his two compatriots. The young men's faces, on the other hand, had angrier looks on them than the priest ever imagined possible.

CHAPTER XX

Not only had his intuition failed him big time, but he'd also sorely misjudged the two brothers. And they looked mad enough to wreak havoc! Suffering, as they were, trauma upon trauma, they deserved an apology -- and the sooner, the better! Some people-patching too!

Glancing toward the sheriff, he suggested the brothers be declared beyond suspicion -- which the lawman did, promising further to pursue Annie's case vigorously. "She was one of our own!" he said.

Father John said to the young men, "When you've finished with Mr. Feldspar, please join me for a bite to eat. We'll talk about your aunt's funeral service and I'll also get you lodging, if you haven't done that already."

They hadn't. They didn't look exactly pacified, but he couldn't tell at whom they might still be angry: himself, or one or both the others in the official three-some. But they didn't refuse his offer, which he took as positive.

"I'll be in the front parlor," he said, and stepped from the room, quickly followed by the sheriff and the judge. When they were several rooms away, they huddled briefly. "Now what?" Father John said.

"Not sure," the judge said. "You got anything else, sheriff?"

"You both know pretty much what I do," he said. "But let's talk."

The judge began. "The other tank had to come from somewhere. They said it was delivered later. And the doctor *did* order *three* tanks. If another *was* brought after they left, *they* couldn't have manipulated it. But then, who did? Turning to the sheriff, he said: "Think we can determine if the pharmacy *did* send a third tank?"

"Of course we can," the sheriff responded, his voice treating it as, at best, a poorly put question.

"If so, then there seem to be only a couple of possibilities," the judge continued. "It was contaminated when it arrived or else someone switched tanks within the house."

"Or," Father John chimed in, "it was switched *between* the house and the pharmacy! *Or,* maybe even it's just some ghastly mistake! But, you've got to agree, *all* those seem highly improbable."

"Yes," the sheriff agreed, "but that's all we got right now!" Which didn't raise spirits in the room any.

"Let's meet at your office around two, sheriff," Father John suggested. "*You'll* have something, by then, and maybe one of us will have also had a brainstorm. And I'll have tucked the nephews in, more at peace, I hope, than just now! I'll *try* to explain and apologize."

None of them exuded much confidence, but that's where it stood as the others departed and Father John settled into a comfortable funeral home chair. He was fidgety and troubled by a multiplicity of things. His intuition had failed; his judgment had run ahead of the facts; and he was now feeling sorry for the young men *and* their aunt, who may not have all that many mourners to mark her passing, for all he knew. Besides, now it all seemed a dead-end, whereas only an hour ago it seemed virtually wrapped up. He decided to relegate the matter to what some call the subconscious, but what he knew as the Spirit. He handed it over to the Lord and began to pray. Only later that evening did he begin to feel his own loss at Annie's death.

After soup and salad, he led the brothers to the town's best motel, but not before giving them the promised apology and explanation, *as well as* next morning's Mass time, just in case. Then he went directly to where he always worked through puzzlements.

As he entered the cool, quiet drugstore, he was relieved to see no other customers. Frieda greeted him with a banana split dish aloft and a questioning look.

"No, Frieda. Today, your biggest chocolate malt, thick enough to last a while! I want to sit deep and stay long," he grinned, though he didn't feel very cheerful.

She brought over ice water and said: "Need to get away?"

"Yes -- to think some."

"I understand," she said, and went to mastermind the huge drink, not asking anything more.

He was going to try to keep the suspicious nature of Annie's death to himself. But he *would* discuss the Burger pharmacy. *What had brought the nephews to the doctor who used that drugstore?*

When Frieda returned with the malt and, characteristically, some water for herself, she sat down and waited while he drew his first, long, satisfying sip and savored it: a good twenty seconds! She finally asked: "Want to talk about it?"

"Yes, actually," he said, smacking his lips. "That's *very* nice," he said, looking down at the glass and drawing out his second word before flashing a huge smile. "Where's Fred?"

"Busy with a passel of orders."

After another pause: "Tell me about Doctor Perkins."

She looked puzzled. "Why?"

"Curious," he said.

"What I meant was, what brings that question on? He's a nice enough man; has a large practice, specializing in families and children. That what you're looking for?"

"What else you know about him?" he asked, taking another sip.

"Not from here originally. But then, none of our doctors are any more. Came about fifteen years ago. Already had one child and now has three more."

"Why doesn't he use your pharmacy?"

"Oh, that! He's Methodist and kinda loyal to other Methodists -- discounts and such, I'm told. So, he uses Wes Young in Burger -- unless someone specifically requests otherwise. We get occasional business."

"I was told today he *always* sends people to Burger."

"Not always; but close enough! Most of his patients end up over there. Why? Is that important?"

"I'd think it would be to you. That must be a lot of business."

"Yes. But what can you do! Just one of those things! Best not to get worked up over stuff like that, you know."

"Guess I'm just glad it's not bad blood between you and him."

"How'd this come up, anyway?"

"Well, just today I learned that's who supplied Annie Verden's oxygen"

"Annie's gettin oxygen? Why?"

"It's a long story, but more to the point, she *was*. No longer! Annie died yesterday." He had blurted it out, despite earlier intentions.

Frieda's face showed complete surprise. Not only was it shocking news, but *a day ago already!* She couldn't imagine not having heard. And then she said as much.

"Well, actually, they really aren't sure what caused it, so they're checking. By now, I suppose, they're letting it out, because her nephews finally got here. But -- and this is *just between us --* they're not sure it was natural causes. *For goodness sake,* don't spread that or they'll know where you got it!" He instantly regretted having let *that* out of the bag too, and desperately hoped Frieda would indeed be discreet.

"My lips are sealed! What do they think, then?"

"Don't know yet. But, truth to tell, I'm more concerned about that oxygen. Her nephews started her on it, and it seemed harmless enough. Still does, I guess. What puzzled me was the *Burger* part."

"Well, as I said," Frieda explained, "it's religious, of all things!"

"Sounds dumb," Father John responded. "Who's this Methodist that's so important? The pharmacist, I mean."

"Oddly enough, he's from Algoma. And *still* lives here!"

"So, why's his business in Burger?"

"That's the religious thing again! Long time ago -- way before we got into business here -- there were two drugstores in town: this one and the Young's across the square. They were apparently overcharging Catholics -- the Germans, actually -- and the Lutherans sided with their Catholic friends. Never was really proved, but it affected the Young's business, enough so's they went to Burger. Wasn't no other druggist, and with the hospital there, it worked out fine financially. But ever since, they haven't done much around here. They'd have *lived* over there too, but no one'd buy their house, so they stayed on here. But they don't take part in *nothin* in Algoma. All civic activity's in Burger. 'Cep maybe votin, I s'pose."

"How long ago was that?"

"Goodness! Eons ago! Wes' daddy made the move, and Wes is old himself now. No kids, though, so I don't know what'll happen to the business when he's gone."

"Old? How old? And doesn't he have anyone with him?"

"Yeah, one younger guy, but he's just hired help. Don't think he's got a part of the business. As for Wes, he *must* be in his eighties."

Father John made a mental note to ask Bob Lanner about the original incident. And then he settled back to enjoy a bit more of the malt. "This is quite nice, Frieda. Thanks again."

"When's the funeral?"

"Two days from now. The nephews need time to plan, and getting word around will take today and tomorrow."

Someone walked in, and Frieda went up to serve up a Coke. Father John wasn't anywhere near comfortable with these new complications, so he finished his drink rather quickly and headed back to the rectory before Frieda could return to his booth.

Before he left the downtown, he decided to stop for that conversation with Mr. Lanner. *Might turn up something helpful for two o'clock.* He saw Maisie dusting when he walked in, so he nodded and made his way over to her. "How are you now, Maisie?" he asked, gently.

"Feelin a little better; had time to pray some for Miss Annie."

"Yes, things like that take time. Oh, while I'm thinking of it, did Miss Annie's oxygen tanks all arrive at the same time?"

"Nope. Two come first and den a few days later one more. Miss Annie told me that, cuz there it sat. She had me move it upstairs. An I did, up the back stairs on that 'traption."

"The chair lift?"

"Yeah. Put it on my lap and took it on up."

"Couldn't the nephews have done that?" he asked, slyly.

"They was done gone by then."

"Oh," he said. "I see. You know who brought it past the house?"

"Nope. It was jes there when I showed up."

"Well, nice to see you, Maisie. After Annie's funeral we'll talk about my rectory. Ok?"

"Yeah. Thank ya, rev'ren. When they buryin Miss Annie?"

"Oh, sorry! Day after tomorrow -- ten am!" He continued on his way to Bob Lanner's office, saying goodbye over his shoulder.

He got in almost immediately, and upon entering, greeted the bank president warmly. Mr. Lanner was equally cordial: "I understand a big front's comin our way. Should finally bring better weather."

The priest replied: "Hadn't heard. Good news! Hope it doesn't rain before Annie's burial. She died a day ago, in case you hadn't heard." He realized immediately how abrupt and insensitive that had been.

"I hadn't," the banker said with a surprised look. "I am *so* sorry to hear that." His look of surprise changed quickly to genuine sadness, and the priest reached across the desk to hold the man's hands gently. And to apologize for not breaking the news more gently.

After a few moments, Father John said: "She'd been using oxygen thanks to her nephews, and I discovered it came from *Burger.* When I asked Fred and Frieda about it, they said Doctor Perkins sends his prescriptions there –- the nephews went to him for some reason. Luck of the draw, I guess. Anyway, the oxygen was from Burger. And I was wondering what caused the rift Frieda told me sent the Youngs over there in the first place. Overcharging or something?"

"Oxygen, you say. Hadn't heard that! Yeah, the Youngs was here when I's a boy. The Germans got miffed, so they skeedaddled in pretty short order, and never looked back. Old man Young did that. I knew his son pretty well, at the time. We was in school together, Wes and me. But we haven't had much contact since. At one time we was rivals over Annie -- in high school. Goodness, that's a while ago!"

"So he dated Annie too! She was a popular girl, then, I guess."

"Oh my, yes! But only a couple fellas ever dated her. Wes and me, and two other guys who are both gone now."

"Rivals, you say. *Really?* Or just a figure of speech?"

"Well, he always took it hard whenever I'd get a date. Kinda jealous, you know. Nothin ever come of it 'tween us. Not even angry words. But no mistakin his displeasure! Kinda funny now; but not then!"

"So it *was* true: the old man was gouging people?"

"Purt near ever'one thought so. Wasn't never proved, like in court, ya know! But didn't matter. The writin was on the wall, and he hightailed it. Never admitted nothin. And never paid anythin back, neither. They've had a *good* business in Burger. Didn't seem fair, specially to the Germans, *none* of whom ever used Young's again! But nowadays the kids don't remember. *They* go there, especially if someone like Perkins sends 'em."

"So, you think it was true, then?"

"No doubt in *my* mind -- or my daddy's. He wouldn't give 'em the start-up loan. Had to get that in Burger. Don't suppose that helped relations 'tween me and Wes any, neither." He laughed quietly.

"Know anything else about him or his business, Mr. Lanner?"

"No, 'ceptin he seems to've made a bunch of money."

"Well, thank you very much, then. Wish I had better news. We'll be burying Annie from our church day after tomorrow at 10 o'clock."

"Thank you, Father. You know I'll be there," he said, as he walked the priest to the door of his office.

CHAPTER XXI

No brainstorms at the rectory. By two o'clock Father John was eager to see the judge and sheriff and was hoping *they'd* have something. What little he'd found out about Wes Young, he feared, wouldn't be much help.

As it turned out, the pharmacist was the topic of conversation. Even before Father John could mention Maisie's contention that the third canister *was* delivered later, the sheriff said he wanted to know more about the oxygen delivery. He had called Mr. Young, who had agreed to come -- reluctantly, it seemed. He was to arrive momentarily, so they huddled only to discover no one had anything else. They could only fruitlessly mull over ideas about the third tank till the pharmacist arrived.

The moment he walked in, Father John had an epiphany. In a flash he was sure he knew the crime's motive and the killer. He just had to verify his intuitions. He listened impatiently as the elderly pharmacist spoke, new insights dancing through his mind the whole time.

In the fifteen-minute conversation the druggist insisted he had sent all three canisters to Algoma the same day, even producing the lading receipt signed by Annie to back that up. Despite the sheriff's thanks, he departed noticeably grumpy, mumbling about wasted time.

To the other two it was another impasse. But not to Father John, who kept his own counsel. He now knew most of what he thought could settle the case. He also thought it was more than coincidence the old man had brought along that receipt. It might seem like convincing evidence, but he believed the matter bore further looking into, convinced as he was that things weren't all they seemed in that regard. But everything in good time!

Father John took his leave and went right back to Bob Lanner. "I just met Wes Young," he said, "and I'd like to ask you a few more questions about him."

"Anythin to help!" the banker said, slightly bemused.

"Well, for starters, what he was like physically as a young man?"

"Physical? You mean, height, weight; like that?"

"Anything that comes to mind when you hear the word."

"Well, he was a little shorter than us others; had unruly hair, I remember; wasn't very athletic, in large part because of his limp …"

"Limp? Do you know what might have caused that?"

"Don't know. Always had it, if I 'member right."

"Broken leg sometime; polio; congenital?"

"Don't rightly know, as I said. But he always had it. He was a class ahead of me, but even in grade school he had a little limp."

"Which leg: right or left?"

"Left. Why?

"Well, he had more than a limp today; and it *was* on his left side. Whatever it was, *now* it's pronounced. Are you sure it wasn't always that way? Significant, obvious, I mean?"

"Not that I recall. All's I member is that he couldn't play sports, and so was very bookish, maybe to a fault. Oh yeah: and he loved science. Came by that honestly, no doubt, what with a pharmacist for a father. It was no surprise when he went to pharmacy school in St. Louis. Though, as I recall, he didn't go right out of high school. Worked at his dad's place a while first."

"Did he go the same year Annie was away, perhaps?"

"That's a strange question, Father. Suppose that could've been the case." After a pause, he said: "Yes, now I think on it, that was exactly so. Whatever made you ask that?"

"Oh, I don't know. I guess Annie's just on my mind a lot these past couple of days," Father John answered evasively.

But it was no idle question. Though Horace was much more thickly built, Wes Young's gait and gimpy physical demeanor reminded the priest strikingly of the junkman. And that suggested a highly believable motive for the old man's possible involvement in Annie's death. It seemed likelier and likelier to the priest that Wes Young, not Robert Lanner, had fathered Annie's bastard.

And while he wasn't going to suggest that to the banker -- not yet, at least, if ever -- he might have to bring it to the attention of the sheriff. But at that precise moment he wasn't sure just how. Not without breaking a confidence! He left the bank as soon as he could after some further small talk concerning the funeral.

As he retraced his route to the county jail, he was thinking about how to engage the sheriff. He almost backed down in front of the jail. But then it suddenly occurred to him how to do it, and he stepped inside the jail, taking a very deep breath for good measure. He wouldn't have to mention Bob Lanner at all!

CHAPTER XXII

"Back already? I thought we weren't meeting till I called everybody," the sheriff said, surprised, not bothering to rise from behind his cluttered desk.

"Got a brainstorm!" Father John could hardly contain himself. "You know, Maisie said the third container definitely came later. Is there any way to find out who does Young's delivering? Something doesn't smell right, and I'm betting if we get to that delivery person quickly, we can find out what really happened. I mean, maybe Mr. Young isn't really on top of that; or forgot. Whatever! In any event, the delivery fellow would know for sure. Annie could have signed that receipt with *verbal* assurances the third one would come soon -- right, sheriff?"

"Well, now you mention it, that *could* be. Let me do some quick and discreet checkin. Maybe we can find out who that delivery boy is."

A phone call to the police in Burger gave him what he wanted to know; and within the hour the young man was sitting in the sheriff's office. He was a nice looking high school senior who easily gave the information about the deliveries -- for there had indeed been more than one, a couple of days apart. He was dismissed and told to keep this conversation strictly to himself for the time being. He looked scared enough to obey.

As soon as he was gone, Father John spoke up. "Sheriff, I have independent reason to believe Mr. Young knows more than he let on. But I'm afraid I can't tell you why I think so. I mean, confidentiality! Please trust me! I'd like to talk to Mr. Young myself. If I'm lucky, perhaps I can get to the bottom of this. You'll still have the opportunity to talk to him yourself, of course, whether my hunch is right or not." The lawman agreed.

"If you don't hear from me within an hour, you can seek out Mr. Young yourself and get to the bottom of that second delivery. I'll call you if I can't locate him. Even if I *can,* I hope to get back to you no more than an hour from now." And with that, Father John set out for Burger.

The Young pharmacy wasn't as inviting as Fred and Frieda's, in part because it was strictly a pharmacy: no soda fountain, just aisles of drugs and over-the-counter stuff! It was also smaller. As Bob Lanner had indicated, another man was working there, but Father John hoped Mr. Young might be there too. He didn't want to leave a message and arouse any suspicion. As hoped, Mr. Young was indeed there and soon appeared up front.

"Mr. Young, you may remember me from the sheriff's office. I don't think we've ever really met before, so permit me to introduce myself. I'm Father John Wintermann from St. Helena's in Algoma."

"Good afternoon," the older man said noncommittally.

"Is there somewhere we could talk privately?"

Mr. Young turned wordlessly and led him to what was apparently his office in the rear. He didn't seem enthusiastic, and was certainly wary, if also curious, his left hand betraying a slight twitch the priest hadn't noticed before.

When they were seated, Father John began: "First of all, in case you might wish to know, we'll bury Annie Verden the day after tomorrow at ten am from my church. I thought you'd want to know, because I understand you and Annie were old acquaintances."

The older man looked startled, and timidly asked, "What do you mean?"

"Well, I understand that you and she dated at one time."

"On occasion. But that was years ago. Sixty or so, I guess."

"Yes, I know. But I have reason to believe you and she were a lot closer than you might be implying."

"And what leads you to say that?" he said a little testily.

"Horace Denver," was all Father John said.

The other man turned ashen, and was about to speak, when Father John added: "No need to say anything foolish, Mr. Young -- like trying to deny what we both know to be true. You see, I know all about Horace's background."

The older man lost even the appearance of any confidence, and perhaps near to tears, put his head into his hands.

"There's no need -- nor do I intend -- to divulge any of this to the public, however. It's in the best interest of a lot of people not to, so I certainly won't let that information out, though I surely could have before now! *But,* Annie's death is another matter. You see, the sheriff now knows there were *two* deliveries, and it will be simple to tie *you* to that third tank. What he'll want to know is who's responsible for its contents. In other words, he'll want to know if that tank was labeled *or* filled by mistake -- or on purpose."

He paused, both to let that sink in and also to watch the other man's face carefully, for the pharmacist had now raised his head and was looking him straight in the eye. Absent any surprise on the older man's face, Father John knew the pharmacist realized exactly what he was implying. He couldn't tell if the look he was now receiving was defiant or not, but it was certainly intense. He decided to continue.

"As a priest, Mr. Young, I'm often bound to confidentiality. And I bind myself to it in this conversation. As a priest, I also need to often distinguish between indiscretions, sins and crimes. Indiscretions can and should be overlooked. Sins are ultimately God's concern. Crimes are public and punishable by the public -- and in very public fashion, at that."

"Now, I happen to believe a crime has been committed. But, as a priest, that's not my concern. My concern is with sin. And if, in fact, a sin has been committed, it's my office, my solemn trust, to facilitate its forgiveness -- and in very confidential fashion, as I implied earlier. What's said here, Mr. Young may not be -- will not be -- repeated by me! Not to anyone, anywhere!" The older man kept staring, noncommittally.

"I think you and I both know why you might have been tempted to take revenge on Annie. When an opportunity you must have thought would never appear suddenly *did* come your way, I can understand how it might have seemed impossible to pass up. That doesn't, of course, excuse anything; but it makes it easier for someone like me to understand." He paused again. Still the other man continued to stare.

"It must have been galling to be spurned by Annie those many years ago. But surely you must have eventually realized that she never took *anyone* into her heart, not even her own child -- not completely -- not openly, at any rate. Her affection for him was distanced and nuanced. To this day I'm not sure he even knows she was his mother!"

"But certainly, she never admitted anyone into a loving relationship, especially not one like you sought so many years ago. If anything, she kept *everyone* at an emotional distance, in what became an aloof, austere role she played all her adult life. It was also, I believe, at base a lonely role. And, I think after a while reality and the role became so entwined as to be probably inseparable, maybe indistinguishable, even by Annie herself. The resentment I believe you've harbored all these years, therefore, has been severely misplaced, Mr. Young. Unfortunately! Especially in light of what has now transpired."

"To make a long story short, Mr. Young, the sheriff is going to have some serious questions for you. As to the truth of your answers, that's between you and God. Only you and God know whether what happened was deliberate or accidental. And what you tell the sheriff, or maybe even some court, is not my concern. It *may or may not* affect your mortal life. But what you may choose to confide *here* -- or *hide – almost certainly* affects your immortal soul. Personally, *if I were you* and had to choose between the two, I'd be more worried about the latter. Especially, if I might point it out, at your age!"

This time when he paused, Father John decided to remain silent to force the other man into speech. It took more than a few moments.

"Father Wintermann," the pharmacist finally began with a deliberate edge to his voice, "Can I trust you when you say this conversation will remain private -- just between us two?"

"You have my personal and professional word, Mr. Young."

The older man looked down silently a long while. When he looked up again, the priest could see anguish on his face. "Why should I say *anything* to you? It may well be twisted or somehow presented to my detriment!"

"Mister Young! I've promised my silence!"

"And why should I believe you?"

"Because I could well have shared this with the authorities already -- and have not! Surely you realize the truth *and* the implications of that. And you must also realize that I could well have a deep concern for justice with regard to my parishioner, Annie Verden. In fact, that is the case. I very much *do* -- and quite earnestly, at that -- want that justice. But *not* at the expense of withholding from you a chance for forgiveness!"

That seemed not only to have quieted the pharmacist, but from the look on his face, surprised him as well.

"Perhaps, given your Protestantism," the priest continued, "you're less familiar with -- and for all I know, less trusting of -- a recognizable human process for forgiving sin, but let me assure you that we Catholics are not only aware of it, but we're comfortable with it, and comforted by it, too. And I really do want to extend that comfort to you -- in large part, because I've guessed you're much in need of it. If that's presumptuous, I apologize. But that's my assumption, and it's the real reason I'm here today."

The old man took his time, looking away the longest while. The priest kept his silence, giving the older man all the time he needed.

When at last he did speak, the old man's voice was noticeably different. He was quiet and subdued, slow and deliberate in his speech. "You're right. I find any so-called human process for forgiving sin strange. Other, that is, than a direct apology between people. And I don't know what to think about what you propose. But I *do* feel the need for forgiveness." After another long pause, he added: "I'll take you up on your offer."

What followed took the better part of twenty minutes, and was never later mentioned by Father John. All he said shortly afterwards to the sheriff was that Mr. Young would welcome a call.

The lawman wasted no time, and within the hour Mr. Young was being questioned again. By day's end, due legal procedure had named the Burger pharmacist a suspect in Annie's death. He was released on his own recognizance as presenting no risk of flight or harm to self or others -- at the recommendation of a retired judge and an unnamed clergyman. And a full investigation was announced, with a preliminary hearing tentatively set for later in the month.

Given that information, the public assumed some horrible accident had brought Annie to her death. Given the pharmacist's age, some sort of accidental misstep was easy to believe. And, finally, since Mr. Young was not only no longer in Algoma but was, in fact, gone many years now, his involvement at this stage of the investigation was not a very volatile matter.

At bottom, the nephews felt reassured they were no longer suspects, *and* that *something* was being done about their aunt's wrongful death. The town was in no undue consternation, so the funeral could proceed quietly and reverentially. Father John's people-patching had also brought himself some peace and allowed him to plan a fitting funeral.

And the wheels of justice *were* turning.

CHAPTER XXIII

At the wake next evening, someone told Father John that the weather report promised a massive front heading their way. "Looks like we're going to get a break!" He allowed as how he'd already heard about it and that it would please him immensely, though he wondered to himself if it might disrupt Annie's burial.

The nephews were subdued the entire evening. They received condolences from parishioners and townsfolk -- people they didn't know -- as graciously as they could. Their somber looks were easily mistaken for grief. But grief was only part of it. They were still smarting from the initial suspicion they'd been under. Father John had tried to reassure them but couldn't tell how effective that had been.

Robert Lanner and his wife were there. Bobbi Sue Langley and her husband came as well, as did Fred and Frieda. Even Horace showed up, dressed the same as always. He even stayed fifteen minutes or so to converse with others who came to pay respects. Father John was surprised to see so many people from the town, including some younger ones with their children. Maybe they sensed a piece of history was slipping away.

Pat Kelly walked up to quietly say how much he appreciated the help from some weeks back, and to add that the matter was properly resolved. "It was nothing, Pat," was all Father John said in return. But the lawyer's face indicated that he took that as understatement.

"When's the will to be read, Pat?"

"Tomorrow, after the burial. I thought it best to do it right away, before the nephews left."

"Why didn't you read it today, then?"

"No need to bother them. They probably had other things to worry about. Anyway, I didn't want to complicate their grief any. They don't stand to inherit much, in case you didn't know."

"I suspected that, Pat. But, I don't think that will worry them."

"Well, I couldn't be sure. And waiting one more day won't hurt anything."

"Right enough! Is it to be a small affair?"

"Well, actually, one other reason I've sought you out just now is to make sure you attend. You're specifically invited. Three o'clock at my office!"

"Oh, really? Will there be many others, then?"

"A surprising number, yes. Besides the nephews, Bob Lanner, and Maisie Brown. As well as Horace Denver!"

"Ah, yes. Maisie! I'm glad she's being remembered. And Bob, as her banker, and probably her confidant! But Horace?" he feigned some surprise as he spoke.

"Yes, Horace," was all the lawyer would say, as he quickly begged off to head toward Bob Lanner, who looked as though he and his wife were making for the door.

Over Bob's shoulder, Father John spotted Lumpy Wurtz, dressed as nicely as ever he'd seen him of a church morning. He moved toward him, and complimented him on that just after greeting him.

"Well, it's the least I thought I could do once I learned that my little picture came from Annie."

"Who told you that, Richard?"

"Horace. Well, I up and asked him, I did, when I spotted him in town a week ago. And that's what he told me. Said he got it from Annie. Wonder where she got it! I was meaning to ask her one of these days. Guess I'll never know now."

"I suppose not," was all Father John thought it right to say.

They spent a further moment or two in conversation. And before he began to move toward the Eversmans who had just walked in, Father John secured a date six weeks hence for his fall visit to the farm. "I'll be there in time for an early supper. Let me bring the dessert, okay?"

He reached Julia & Coonie just after they'd signed the guest book. "How are you feeling?" he asked Coonie with some concern.

"Just fine, Father. Jules takes good care of me. She changed my diet -- and the doctor even allows a bit of walking again. Isn't this a shame about Annie? And now they're saying some accident took her!"

"Sounds that way. I'm sure they'll figure it all out. Meanwhile, it's nice you came tonight. Be at the funeral tomorrow?" It was probably a silly question, but given Coonie's recent hospital stay, not totally so.

"You can be sure of it. I'll usher. *Greet!*" he smiled, winking to indicate he'd said it that way on purpose. "Need a reader too?"

Julia looked at him expectantly. "Sure. I can always use the help."

"Count on it, then," she said quietly, and smiled at her priest.

He let them move on toward Annie's casket and the nephews, and turned to survey the room. Soon he'd begin the rosary. He winked across the room at Bobbi Sue and held up his rosary as if to say he expected her participation. She grinned, feigned horror, but stayed nonetheless. Soon the whole room was filled with the quiet, prayerful murmur of those familiar Catholic prayers.

CHAPTER XXIV

The morning of Annie's funeral dawned very cloudy. Weather predictions seemed accurate for once. Temperatures began dropping overnight and the morning sky grew darker by the hour: signs of a front. But rain held off till after the funeral Mass.

Just before nine, Father John heard from the sheriff. Mr. Young had steadfastly maintained ignorance about the contents of the third tank. Without divulging anything he'd learned from the pharmacist, the priest urged an inquiry with the firm that supplied the tank. "They should be able to verify his story."

"Way ahead of you, Father. We're on that. I'll keep you informed. Actually, the judge volunteered to check it out. He knows what to ask and how. But the old man's story about forgetting there were two deliveries is certainly lame."

"I'll wait to hear," the priest said before hanging up.

In his homily, Father John spoke of Annie's contributions to Algoma and St. Helena's throughout her long life. But he also spoke of strange wrinkles in God's Providence, allowing life to touch life in ways seldom predictable, rarely fully understood or even, sometimes, known at all. His delivery was unusually slow and deliberate. While speaking about the people Annie had influenced in her long lifetime, he also encouraged everyone to apply his words to themselves.

"Think, if you will, about those whom you have touched in similar fashion. And, in gratitude, prayerfully remember those who have influenced your life, including Miss Annie Verden," he concluded.

As he looked around, he could see Bob Lanner's eyes tearing up. He also noticed the nephews sitting stoically up front. And there was Horace, his face unreadable; but he was there. And, he noticed, Wes Young was not.

As he rode to the cemetery with the elder Feldspar, conversation was minimal. Larry acknowledged that the more elderly people he buried, the harder it was becoming to face the fact that his own time was coming. And soon enough, no doubt! Father John thought it best to turn the conversation. He mentioned the welcome break in temperature. "But I also think we're going to really get dumped on, and maybe before we're done here."

"Best be quick," said the undertaker, with just the hint of a smile.

"Usually am, right?" the priest smiled back.

As it turned out, the return from the graveside had barely begun for the faithful folks who went to see Annie off, when the heavens opened. Only a few made it safely before the deluge. Father John wasn't one of them. He was soaked, even though he'd gotten his umbrella open quickly. But getting wet didn't bother him. He found himself hoping, rather, that it wouldn't turn into a gully-washer and ruin crops. He was grateful to learn later that it didn't. Had it been sudden -- a thunderstorm, with hail, perhaps -- that might not have been the case. But this front brought a steady rain, with little agricultural harm.

It was still raining at 2:45 when Father John pulled up in front of Pat Kelly's office. He'd picked up the nephews and had just remarked that rain would be a steady diet the rest of the day. They didn't respond, but simply got out and dashed to safety, shaking themselves off inside. Father John took his time before getting out, standard procedure for the more deliberate older man. Then he raised his umbrella and made for the office doorway.

Horace was already there, along with Maisie. Apparently the junkman had thought to offer her a ride. They had only to wait for Bob Lanner. And minutes later when they were all in the waiting room, the secretary received the okay to usher them into Pat Kelly's conference room.

Pat was smiling but business-like as he greeted everyone, his tall, thin frame very erect at the head of a long conference table. "Thank you for coming. Reading the will this quickly accommodates her nephews, who, I am sure, appreciate your thoughtfulness. Before we begin, one person mentioned in the will has declined to appear, for reasons that will become obvious in a moment. That said, let me begin." He droned on through a lot of legalese, including words to the effect that, as her executor, he was entitled to a certain small percentage of the estate's value before any other disposition of assets. But he finally reached the section where Annie's heirs were mentioned.

"As to the disposition of my estate," he quoted, "I leave the following items to the following people. First, to my pastor and friend, Father John Henry Wintermann, I bequeath one painting of his choice -- exclusive of the family portraits -- from the second floor of my home. As he has seen them all, I feel certain he can make his choice with appropriate dispatch. I thank him for his many years of service in Algoma, as well as for his devotion to my spiritual well being over the years." At this point the lawyer looked up to add parenthetically: "This item was recently amended," and he looked directly at Father John.

Then he continued: "To my two great-nephews, the only surviving members of our line, I leave the family portraits, all to be found in one second floor bedroom. I also leave such antiques and heirlooms as they should choose, and designate that their choice be made before the processing of any other provisions impinging upon those choices. The identity of each portrait subject, as well as information about the place of each in the family tree, can be found on the reverse of those upstairs portraits. Such information, I assume, will be of interest to these surviving members of the Verden line."

"Third," the lawyer continued, "I leave to Miss Maisie Brown the contents of the envelope with her name affixed thereto, which envelope is to be found in the strongbox in the kitchen pantry of my house." Mr. Kelly looked up to explain that he knew the location of that strongbox but was unaware of the contents of the envelopes, because: "Miss Verden indicated quite some time ago that she had been putting things into the box for various people for many years. I got the impression that the names grew and dwindled over time, and that the list was changed again recently."

Again he looked down and continued: "Fourth, I wish to thank Mr. Robert Lanner for more years of financial advice than either of us might wish to count, as well as for many, many years of friendship and concern." At that moment Father John realized Bob Lanner had come without his wife. He also realized instantly why.

Pat Kelly was continuing: "Also in that strongbox are several of what I term *books of memory*. There is one with Mr. Robert Lanner's name appended. I bequeath it to him with my fondest thanks, along with an envelope bearing his name. I ask him to follow its instructions, and I thank him for this final service to me and my family."

"Next," the lawyer continued, "is another envelope -- for Mr. Wesley Young. I leave that to the gentleman, who no longer lives in our town."

As Father John noted the absence of any further explanation, and knowing the probable reason, he listened as the lawyer looked up again to say: "Given recent developments, it is no doubt understandable why Mr. Young has declined to appear. I will see that this is delivered to him within the week."

"To Mr. Horace Denver," he continued, "I leave the many remaining paintings on my home's second floor. His indispensable service to me regarding other such paintings is much appreciated, and I designate that he dispose of these paintings as he sees fit and receive any revenues they may generate. Furthermore, I ask him to carefully follow the instructions found in another envelope bearing his name, and to use its contents in any way he may deem fit. There is also a book of memories for him."

"Finally, I ask that my house, along with all contents that remain after the above provisions of my will have been executed, as well as any other remaining assets, be given to Saint Helena's Parish. There is a final envelope addressed to the pastor of Saint Helena's, whose contents explain the few provisions and suggestions I have to help the parish use my home well."

The lawyer read the concluding words of the will, which asked for prayers for the soul of the old lady. Then he added: "I have decided that since there are really no monetarily quantifiable assets, I shall forgo the usual executor's fee. And now, here are the various envelopes to which the will refers. I ask you to take your appropriate envelope -- and you also, Father, because you receive the one addressed to the pastor of Saint Helena's. As I said, I myself shall deliver Mr. Young's." With that he handed them out. After a momentary awkwardness, some began to open theirs.

Maisie was the first to react. She gasped: "Oh, Lordy! So much money!" She was holding five hundred dollars, in crisp twenty-dollar bills.

"Congratulations, Maisie," said Father John, chancing a quick glance at Bob Lanner, who wasn't reacting. The priest wondered if it was no surprise. "I'm so happy for you, Maisie," he said genuinely.

Horace did not open his, but stared down at it instead. The look on his face struck the priest as no more than ordinarily enigmatic. But he was still puzzled at the man's behavior.

Bob Lanner's envelope was far thicker than the others'. Whatever else it contained, he pulled out several sheets of paper and began to read.

Father John had yet to open his own, the one with instructions about the mansion. He had chosen to discreetly look around the office instead.

The nephews were chatting quietly. They seemed neither surprised nor upset -- nor curious about anyone else, either. Maisie still looked thunderstruck. And Bob Lanner was now busy reading a lengthy letter.

Before Father John could attend to his envelope, Mr. Kelly spoke up again. "Having given you all a chance to open the envelopes, I will now give out the memory books mentioned in the will." With that he handed one to Horace and one to Bob Lanner. "There is yet another such book," the lawyer added, handing it to Father John. "It has *Pastor of Saint Helena's* on it, and I assume it has to do with the gift of the Verden home. There was no mention of it in the will, nor was there ever a conversation about it between Miss Verden and myself. In any event, logic dictates that I hand it over now. Nor should there be any legal dispute about this, because it would otherwise fall under the category of *other contents* of the house, and would thus become property of the parish anyway." And he handed the sizeable book to the priest.

"That concludes the disposition of Miss Verden's properties as per the provisions of her will. If there are no further questions of me, I thank you for attending." He looked around, and when no one indicated any wish to speak, he added: "Please feel free to stay as long as you wish. And if you subsequently develop a question or concern, I will be in my office." And with that, he left the room.

CHAPTER XXV

It took a while for things to settle in Algoma. But the cooler weather helped calm passions, and curiosity also. In many people, that is; but not in the pastor of Saint Helena's!

It was true that the nephews were more than satisfied to retrieve their aunt's heirlooms and leave, never to be seen again in Algoma. It is also true that Maisie was overwhelmed with her windfall, but not so much as to give up cleaning the great Algoma homes. And she took on Father John's rectory soon thereafter, a job she would keep as long as he was there.

It was furthermore true that, despite Annie's paintings, Horace Denver continued to work the junkyard, never really changing habits, patterns or appearance. It was to be a long time before townsfolk found out that Annie's paintings finally ended up on the east coast, where they were making a minor splash in the art world. Only Father John learned, eventually, what Horace was doing with the monies from their sale. And he kept his promise not to tell. But he never learned what the junkman's envelope or book of memories contained.

And it was furthermore true that the official investigation dragged on a long while before ending with a slap on the wrist of Wesley Young. It couldn't be satisfactorily determined how the lethal contents of the one tank came to be there, and he was given probation -- because of his advanced age -- after being charged in effect with running his business carelessly: reckless endangerment, or some such charge. He decided -- agreed -- to step down from practice as a pharmacist. His business was closed and liquidated, and he moved away to be with distant relatives. Father John never indicated, not even to the sheriff or judge, whether he thought justice had been done by that court decision; nor did they ask. The official outcome was reported to the nephews. They too were silent about their feelings on the matter.

The mansion was taken over by Saint Helena's. The parish blue ribbon oversight committee decided to honor its historicity, as Annie had more or less suggested in her letter to the pastor. So, with the artifacts that remained after Horace and the nephews had removed their share, they turned the home into a tourist attraction. Modest fees accrued to the parish education fund. Maisie Brown even agreed reluctantly to once again clean biweekly, as well as to consult with the committee in their initial stages of opening the home to the public. Father John and several other townspeople also contributed their collective memories as to the original condition of the building, and Horace was persuaded to loan several paintings on an indefinite basis. Finally, Mr. Robert Lanner established an endowment to provide for upkeep and repair. While numerous townspeople donated to the fund, Father John suspected the bulk of it came from the banker.

Pat Kelly never spoke about the case with Father Wintermann again, either to share his own thoughts or to seek out the priest's. Father John was privately relieved about that.

Bob Lanner did, however, wish to speak with him. The moment occurred well before the sheriff's investigation was fully resolved, two weeks after Annie's funeral. And it took place not in the bank, but at a restaurant in the Saint Louis suburbs, an eatery unfamiliar to Father John, but one where Bob Lanner apparently felt safe from prying ears.

Father John was asked to come in his own car, so he didn't know exactly whom or what to expect. It turned out to be just the banker, who was waiting at a table in the rear of the quiet French bistro. It was 2:30 in the afternoon and, as Father John soon learned, the time and place had been carefully chosen both for secrecy's sake and to allow the banker to return early enough to avoid spousal suspicion. Mr. Lanner wanted everyone he could think of kept in the dark, including Bobbi Sue and the bank staff, as well as his wife. It was why, he explained at the outset, he had insisted they come in separate cars. Father John's curiosity was near to exploding.

"Before anythin else, you want somethin to eat?" the banker asked.

"Too early for my supper, Mr. Lanner," Father John said, his voice trailing off.

"I thought as much too. I'll have to eat at home in a couple a hours, ya know. But we *could* have somethin to drink."

"Possibly. Will we be long here? I mean: I don't want to be driving with much alcohol in my system."

"Nor I. I figure an hour or so. Irish Coffee?"

"Sounds fine."

Once the drinks arrived and both men began to nurse them, the banker said, "Ever'one in Pat Kelly's office that day has kept his own counsel ever since. Haven't been able to pick up a thing round

211

town the past two weeks. Am I right? Have *you* heard anythin?" he asked.

"No, I believe you're correct, Mr. Lanner."

"Well, I'd like to chat 'bout what I received from Annie that day." The priest tried to look calm.

"The envelope held two very int'restin things: a note, and some money. A lot of money, in fact." He paused to look at the priest's face, which told him nothing; so he continued. "All those years I was givin Annie money, she must've been stashin a good bit of it away. I mean: she gave Maisie a bundle. And I'll bet she gave a bunch to Horace. She gave *me* back well over a thousand!" Father John's face finally registered surprise.

The banker continued: "Her note spoke of other things, but as to the money, she told me to use it to help pregnant young girls who didn't know what to do -- who were, ya know, confused or ashamed or goodness knows how else discombobulated."

Father John hesitantly commended the idea, and the banker agreed. "But that ain't my point, Father. I don't know what to do 'bout that. And I's hopin you can help."

Father John said: "I'm sure I can, and will gladly. But we didn't have to come all the way over *here* to talk about that."

He'd struck a nerve -- the banker's face showed as much. "Well, there's more. And I'll get to it soon. But, you can help with *this*?"

"Yes. It's relatively simple. Once I make the contacts, I'll get back to you. But, let me ask: is this supposed to be just for Algoma girls -- or girls in general?"

"She didn't specify."

"So, do *you* wish to -- before I begin to set up anything?"

"Well, s'pose I can," the banker said slowly. "You think it's okay?"

"She put you in charge, didn't she? I think that means you can determine the final look of things."

"Okay, then! I think we'll do it just for Algoma girls. Okay?"

"Fine. But we'll have to get a confidential means to identify them. You trust me to work that out? I have in mind talking to the other pastors plus the town doctors."

"Sounds good!"

"One other matter: do you wish to be connected with it?"

"Oh, my goodness, no," said the horrified banker.

"Fine. It doesn't have to be. In fact, I agree that it's better left in the hands of the clergy and medical community. However, a thousand dollars or two probably won't help much."

"I been wonderin about that too. And I thought I could add a little to it."

"It would have to be a lot, I fear, Mr. Lanner. Why don't we get it going and then let word out to our congregations and the whole town that donations are welcome. With any luck, we'll be able to create and sustain a corpus for ongoing help. You can donate then, if you like." All that remained was the fine-tuning, which Father John promised to do.

"But you said there was another matter or two in the letter."

"*One*," the older man said. But he seemed reluctant to continue.

Father John thought he might not know how to begin -- or maybe didn't want to. "If I'm on eggshells here, I'll back off, Mr. Lanner."

"No, it's not that. It's just -- well, I'm confused by what else she said. Or didn't say! She tippy-toed around the issue of Horace. And frankly, I'm not sure why -- or, for that matter, exactly what she was tryin to say."

Father John felt a lump rise into his throat. He didn't want to get into the issue of Horace's father. To buy time, he said: "Did you happen to bring the letter, Mr. Lanner? If so, do you mind my reading it? What you've said just now isn't very clear, I hope you understand."

"Yes, I did bring it. I was thinkin you might need to see it. In fact, I was wantin you to. I don't mind a'tall," he said, retrieving it from his inner coat pocket and handing it across the table.

Father Wintermann slowly unfolded the three pages. He skipped over a page of thanks for monetary help. Then he also skipped over the handling of the money. Finally, on the last and shortest page, he came to: *Dearest Robert, you can only guess at my gratitude for many years of friendship since our wonderfully remembered school days. Though we never spoke much of these things in so many words, you have been very loyal to Horace and myself. I cannot thank you enough for that. Poor, dear Horace doesn't know about you and me. I have always thought it best not to go into that. In fact, there is much I haven't gone into with him. I'm sure you do understand. And I believe -- and hope -- that you will continue to understand, should I precede you in death. If you are reading this, no doubt that has happened! I trust you'll not share anything with him either. It has all been so complicated these many years! I believe Horace is truly unaware of most of the facts. And I hope he will always be, for I fear he is unable to handle the real truth of it all. In heaven we can all enjoy each other in the truth of eternity. Until which, all my love, my dearest Robert ...*

The priest looked up and realized that the banker had tears in his eyes. "Are you all right?" he asked the older man.

"Yes," he said simply. "I don't recall her ever calling me *dearest* since our youth! But what troubles me most 'bout the letter is a vague feelin. Annie seems to be trying to say somethin more. And I can't figure what; or even if I'm right in the first place 'bout that feelin."

Father John gulped and said tentatively. "I'm not sure what you mean, Bob."

"Well, it's all so vague, I don't know how to speak of it myself. I just think there's more she was tryin to say."

"I'd say you may never know, Bob," said the priest quietly, taking another sip from his drink.

"Well, it couldn't hurt to ask," the banker said softly.

They sat awhile in silence, each tending his drink, before Father John hesitantly said: "I was struck by the fact that she left a note for Wes Young. It's rather ironic, under the circumstances." He watched the banker closely.

"I just thought it was cuz he's her only other suitor still alive. But you're right, it *is* ironic." The banker's face and demeanor gave no hint of understanding anything more than exactly what he'd just mentioned. And Father John, relieved, left it at that.

Within another fifteen minutes both men were in their cars. Father John left just moments ahead of the banker and, just beating a stoplight, was able to pull well ahead in traffic. They'd arrive back in Algoma at very different times. That afternoon's restaurant moment effectively ended any more conversation on the matter between them. And, again, Father John was thankful.

He was uncharacteristically satisfied that life in Algoma had returned to its usual sleepy self. Fred and Frieda were starved for gossip, the summer festivals were over, with only the Methodist Trough remaining more than a month hence. And there was a rumor they were re-thinking their decision to change from serving their time-tested Mulligan. People were back in church after summer vacations, no more or fewer than before summer started. Summer with its high temperatures was, in fact, a thing of memory. Labor Day effectively ended it, even if meteorologists might wait several weeks to declare it formally finished. Church matters were routine again, and Father John was not unhappy over it. For another thing, the change of seasons had brought cooler weather, something he welcomed whenever and however he might find it. The unrelenting heat had been most unkind that year, and it had helped rob them of Annie.

For a moment he thought of the pitiless heat as not unlike Annie, until he realized that was unfair -- and for many reasons. There was her final testament's generosity and sensitivity to numerous people. There was also Horace, who, incidentally, had confided to him one day near September's end that Annie had been almost like a mother to him. Father John hadn't known what to say in response, and so said nothing.

Annie and the weather weren't alike after all, he decided, but the town's memory of that hot summer would fade and the heat be unremembered, much like Annie herself, he feared. Although, there'd be, no doubt, an Indian Summer to remind them of those warm months. Perhaps Horace would serve that function with regard to Annie -- at least for Bob Lanner and himself, and maybe even Wes Young, for all he knew!

He was content with that: Horace as *Annie's* Indian Summer -- and probably as unaware as any Indian Summer itself was of its own seasonal role. So be it! It signaled the end of something that needed ending. And the beginning of something new, whatever that might be. In the end, he was happy Horace would play his part in that grand scheme.

And there were things to look forward to: Lafe Skinner's flowers and Richard Wurtz's farm. Life would go on.

To order copies of "Heat", please fill out the order and shipping forms and enclose them + your check in an envelope addressed to:

PAX Publications

7315 Henderson CT SE

Olympia WA 98501

Inquiries may be sent to the above address or to:

jackfrerk@aol.com

- - - - - - - - - -

ORDER FORM

Please ship _____ copies of **HEAT** by Jack Frerker

Retail price: $13.00 (plus $2.00 per copy Shipping & Handling)

Phone: _____ E-mail: _____

Make checks payable to: PAX PUBLICATIONS

Payment of $_____ is enclosed

Signature _____

Shipping Information:

Name: _____

Address: _____

City: _____

State: _____ Zip: _____